COMMON NAMES OF
MAMMALS OF THE WORLD

G000162861

COMMON NAMES OF MAMMALS OF THE WORLD

DON E. WILSON AND F. RUSSELL COLE

WITH CONTRIBUTIONS BY
BERNADETTE N. GRAHAM, ADAM P. POTTER,
AND MARIANA M. UPMEYER

SMITHSONIAN INSTITUTION PRESS
Washington and London

© 2000 by the Smithsonian Institution
All rights reserved

Editorial proofreader and indexer: Barry R. Koffler
Production editor: Duke Johns
Designer: Janice Wheeler

Library of Congress Cataloging-in-Publication Data
Wilson, Don E.
    Common names of mammals of the world / Don E. Wilson and F. Russell Cole.
      p.   cm.
    Includes bibliographical references (p.   ).
    ISBN 1-56098-383-3 (pbk.: alk. paper)
     1. Mammals—Nomenclature (Popular).   I. Cole, F. Russell.   II. Title.
QL708.W45   2000
599'.01'4—dc21   99-087359

British Library Cataloguing-in-Publication Data available

Manufactured in the United States of America
07 06 05 04 03 02 01 00 5 4 3 2 1

∞ The paper used in this publication meets the minimum requirements of the American
National Standard for Information Sciences—Permanence of Paper for Printed Library Materials
ANSI Z39.48-1984.

CONTENTS

1   Introduction

7   ORDER MONOTREMATA (Monotremes)
7   Family Tachyglossidae (Echidnas)
7   Family Ornithorhynchidae (Platypus)

7   ORDER DIDELPHIMORPHIA (American Opossums)
7   Family Didelphidae
7   Subfamily Caluromyinae
7   Subfamily Didelphinae

10   ORDER PAUCITUBERCULATA (Shrew Opossums)
10   Family Caenolestidae

10   ORDER MICROBIOTHERIA (Monito Del Monte)
10   Family Microbiotheriidae

11   ORDER DASYUROMORPHIA (Carnivorous Marsupials)
11   Family Thylacinidae (Thylacine)
11   Family Myrmecobiidae (Numbat)
11   Family Dasyuridae (Dasyurids)

13   ORDER PERAMELEMORPHIA (Bandicoots and Bilbies)
13   Family Peramelidae (Australian Bandicoots and Bilbies)
13   Family Peroryctidae (Rainforest Bandicoots)

14   ORDER NOTORYCTEMORPHIA (Marsupial Moles)

14   Family Notoryctidae

14   ORDER DIPROTODONTIA (Koalas, Wombats, Possums, Macropods, and relatives)

14   Family Phascolarctidae (Koala)
14   Family Vombatidae (Wombats)
14   Family Phalangeridae (Cuscuses and Brushtail Possums)
15   Family Potoroidae (Bettongs, Rat Kangaroos, and Potoroos)
15   Family Macropodidae (Kangaroos and Wallabies)
17   Family Burramyidae (Pygmy Possums)
17   Family Pseudocheiridae (Ringtail Possums)
18   Family Petauridae (Gliding and Striped Possums)
18   Family Tarsipedidae (Honey Possum)
18   Family Acrobatidae (Feathertail Gliders and Possums)

18   ORDER XENARTHRA (Edentates)

18   Family Bradypodidae (Three-toed Sloths)
19   Family Megalonychidae (Two-toed Sloths)
19   Family Dasypodidae (Armadillos)
19   Subfamily Chlamyphorinae
19   Subfamily Dasypodinae
20   Family Myrmecophagidae (American Anteaters)

20   ORDER INSECTIVORA (Insectivores)

20   Family Solenodontidae (Solenodons)
20   Family Nesophontidae (Nesophontes)
20   Family Tenrecidae (Tenrecs)
20   Subfamily Geogalinae
21   Subfamily Oryzorictinae
21   Subfamily Potamogalinae
21   Subfamily Tenrecinae
21   Family Chrysochloridae (Golden Moles)
22   Family Erinaceidae (Hedgehogs and Moonrats)
22   Subfamily Erinaceinae
23   Subfamily Hylomyinae

23   Family Soricidae (Shrews)

23   Subfamily Crocidurinae

29   Subfamily Soricinae

32   Family Talpidae (Desmans, Moles, and Shrew Moles)

32   Subfamily Desmaninae

32   Subfamily Talpinae

34   Subfamily Uropsilinae

34   ORDER SCANDENTIA (Tree Shrews)

34   Family Tupaiidae

34   Subfamily Tupaiinae

34   Subfamily Ptilocercinae

35   ORDER DERMOPTERA (Flying Lemurs)

35   Family Cynocephalidae

35   ORDER CHIROPTERA (Bats)

35   Family Pteropodidae (Old World Fruit Bats)

35   Subfamily Pteropodinae

41   Subfamily Macroglossinae

41   Family Rhinopomatidae (Mouse-tailed Bats)

41   Family Craseonycteridae (Hog-nosed Bat)

41   Family Emballonuridae (Sheath-tailed Bats)

43   Family Nycteridae (Slit-faced Bats)

43   Family Megadermatidae (False Vampire Bats)

44   Family Rhinolophidae (Horseshoe Bats)

44   Subfamily Rhinolophinae

45   Subfamily Hipposiderinae

48   Family Noctilionidae (Bulldog Bats)

48   Family Mormoopidae (Leaf-chinned Bats)

48   Family Phyllostomidae (American Leaf-nosed Bats)

48   Subfamily Phyllostominae

49   Subfamily Lonchophyllinae

50   Subfamily Brachyphyllinae

50   Subfamily Phyllonycterinae

50   Subfamily Glossophaginae
51   Subfamily Carolliinae
51   Subfamily Stenodermatinae
53   Subfamily Desmodontinae

54   Family Natalidae (Funnel-eared Bats)
54   Family Furipteridae (Thumbless Bats)
54   Family Thyropteridae (Disk-winged Bats)
54   Family Myzopodidae (Sucker-footed Bat)
54   Family Vespertilionidae (Vesper Bats)

54   Subfamily Kerivoulinae
55   Subfamily Vespertilioninae
63   Subfamily Murininae
63   Subfamily Miniopterinae
63   Subfamily Tomopeatinae

64   Family Mystacinidae (New Zealand Short-tailed Bats)
64   Family Molossidae (Free-tailed Bats)

66   ORDER PRIMATES (Primates)

66   Family Cheirogaleidae (Dwarf Lemurs and Mouse Lemurs)

66   Subfamily Cheirogalinae
67   Subfamily Phanerinae

67   Family Lemuridae (Large Lemurs)
67   Family Megaladapidae (Sportive Lemurs)
67   Family Indridae (Leaping Lemurs)
68   Family Daubentoniidae (Aye-aye)
68   Family Loridae (Angwantibos, Lorises, and Pottos)
68   Family Galagonidae (Galagos)
68   Family Tarsiidae (Tarsiers)
69   Family Callitrichidae (Marmosets and Tamarins)
70   Family Cebidae (New World Monkeys)

70   Subfamily Alouattinae
70   Subfamily Aotinae
70   Subfamily Atelinae
71   Subfamily Callicebinae
71   Subfamily Cebinae

72   Subfamily Pitheciinae

72   Family Cercopithecidae (Old World Monkeys)

72   Subfamily Cercopithecinae

73   Subfamily Colobinae

75   Family Hylobatidae (Gibbons)

75   Family Hominidae (Humans and Great Apes)

75   ORDER CARNIVORA (Carnivores)

75   Family Canidae (Dogs)

76   Family Felidae (Cats)

76   Subfamily Acinonychinae

77   Subfamily Felinae

78   Subfamily Pantherinae

78   Family Herpestidae (Mongooses)

78   Subfamily Galidiinae

78   Subfamily Herpestinae

79   Family Hyaenidae (Hyenas)

79   Subfamily Hyaeninae

80   Subfamily Protelinae

80   Family Mustelidae (Weasels, Badgers, Skunks, and Otters)

80   Subfamily Lutrinae

80   Subfamily Melinae

81   Subfamily Mellivorinae

81   Subfamily Mephitinae

81   Subfamily Mustelinae

82   Subfamily Taxidiinae

82   Family Odobenidae (Walrus)

82   Family Otariidae (Sea Lions)

83   Family Phocidae (Earless Seals)

84   Family Procyonidae (Raccoons and relatives)

84   Subfamily Potosinae

84   Subfamily Procyoninae

84   Family Ursidae (Bears)

84   Subfamily Ailurinae

84   Subfamily Ursinae

85   Family Viverridae (Civets and relatives)

85   Subfamily Cryptoproctinae

85   Subfamily Euplerinae

85   Subfamily Hemigalinae

85   Subfamily Nandiniinae

85   Subfamily Paradoxurinae

86   Subfamily Viverrinae

86   ORDER CETACEA (Whales, Dolphins, and Porpoises)

86   Family Balaenidae (Right Whales)

86   Family Balaenopteridae (Rorquals)

87   Family Eschrichtiidae (Gray Whale)

87   Family Neobalaenidae (Pygmy Right Whale)

87   Family Delphinidae (Marine Dolphins)

88   Family Monodontidae (White Whales)

88   Family Phocoenidae (Porpoises)

89   Family Physeteridae (Sperm Whales)

89   Family Platanistidae (River Dolphins)

89   Family Ziphiidae (Beaked Whales)

90   ORDER SIRENIA (Manatees, Dugongs, and Sea Cows)

90   Family Dugongidae (Dugongs and Sea Cows)

90   Family Trichechidae (Manatees)

90   ORDER PROBOSCIDEA (Elephants)

90   Family Elephantidae

90   ORDER PERISSODACTYLA (Odd-toed Ungulates)

90   Family Equidae (Horses, Zebras, and Asses)

91   Family Tapiridae (Tapirs)

91   Family Rhinocerotidae (Rhinoceroses)

91   ORDER HYRACOIDEA (Hyraxes)

91   Family Procaviidae

91 ORDER TUBULIDENTATA (Aardvark)

91 Family Orycteropodidae

92 ORDER ARTIODACTYLA (Even-toed Ungulates)

92 Family Suidae (Pigs and Hogs)
92 Subfamily Babyrousinae
92 Subfamily Phacochoerinae
92 Subfamily Suinae

92 Family Tayassuidae (Peccaries)
92 Family Hippopotamidae (Hippopotamuses)
93 Family Camelidae (Camels and relatives)
93 Family Tragulidae (Chevrotains)
93 Family Giraffidae (Giraffes and Okapi)
93 Family Moschidae (Musk Deer)
93 Family Cervidae (Deer)

93 Subfamily Cervinae
94 Subfamily Hydropotinae
94 Subfamily Muntiacinae
94 Subfamily Capreolinae

95 Family Antilocapridae (Pronghorn)
95 Family Bovidae (Cattle, Antelopes, Sheep, and Goats)

95 Subfamily Aepycerotinae
95 Subfamily Alcelaphinae
95 Subfamily Antilopinae
97 Subfamily Bovinae
98 Subfamily Caprinae
99 Subfamily Cephalophinae
99 Subfamily Hippotraginae
100 Subfamily Peleinae
100 Subfamily Reduncinae

100 ORDER PHOLIDOTA (Pangolins)

100 Family Manidae

100 ORDER RODENTIA (Rodents)

100 Family Aplodontidae (Mountain Beaver)

100 Family Sciuridae (Squirrels)

100 Subfamily Sciurinae

108 Subfamily Pteromyinae

110 Family Castoridae (Beavers)

110 Family Geomyidae (Pocket Gophers)

111 Family Heteromyidae (Pocket Mice, Kangaroo Rats, and Kangaroo Mice)

111 Subfamily Dipodomyinae

112 Subfamily Heteromyinae

112 Subfamily Perognathinae

113 Family Dipodidae (Jerboas)

113 Subfamily Allactaginae

114 Subfamily Cardiocraniinae

114 Subfamily Dipodinae

114 Subfamily Euchoreutinae

114 Subfamily Paradipodinae

114 Subfamily Sicistinae

115 Subfamily Zapodinae

115 Family Muridae (Rats, Mice, Voles, Gerbils, Hamsters, and Lemmings)

115 Subfamily Arvicolinae

120 Subfamily Calomyscinae

120 Subfamily Cricetinae

121 Subfamily Cricetomyinae

121 Subfamily Dendromurinae

122 Subfamily Gerbillinae

125 Subfamily Lophiomyinae

125 Subfamily Murinae

144 Subfamily Myospalacinae

144 Subfamily Mystromyinae

144 Subfamily Nesomyinae

145 Subfamily Otomyinae

145 Subfamily Petromyscinae

145 Subfamily Platacanthomyinae

146 Subfamily Rhizomyinae

146 Subfamily Sigmodontinae

160   Subfamily Spalacinae

160   Family Anomaluridae (Scaly-tailed Squirrels)

160   Subfamily Anomalurinae

160   Subfamily Zenkerellinae

161   Family Pedetidae (Spring Hare)

161   Family Ctenodactylidae (Gundis)

161   Family Myoxidae (Dormice)

161   Subfamily Graphiurinae

161   Subfamily Leithiinae

162   Subfamily Myoxinae

162   Family Bathyergidae (Blesmols)

162   Family Hystricidae (Old World Porcupines)

163   Family Petromuridae (Dassie Rat)

163   Family Thryonomyidae (Cane Rats)

163   Family Erethizontidae (New World Porcupines)

164   Family Chinchillidae (Viscachas and Chinchillas)

164   Family Dinomyidae (Pacarana)

164   Family Caviidae (Guinea Pigs)

164   Subfamily Caviinae

164   Subfamily Dolichotinae

165   Family Hydrochaeridae (Capybara)

165   Family Dasyproctidae (Agoutis)

165   Family Agoutidae (Pacas)

165   Family Ctenomyidae (Tuco-tucos)

166   Family Octodontidae (Octodonts)

167   Family Abrocomidae (Chinchilla Rats)

167   Family Echimyidae (American Spiny Rats)

167   Subfamily Chaetomyinae

167   Subfamily Dactylomyinae

167   Subfamily Echimyinae

168   Subfamily Eumysopinae

169   Subfamily Heteropsomyinae

170   Family Capromyidae (Hutias)

170   Subfamily Capromyinae

170   Subfamily Hexolobodontinae

170   Subfamily Isolobodontinae
170   Subfamily Plagiodontinae

170   Family Heptaxodontidae (Key Mice)
170   Subfamily Clidomyinae
171   Subfamily Heptaxodontinae

171   Family Myocastoridae (Nutria)

171   ORDER LAGOMORPHA (Rabbits, Hares, and Pikas)

171   Family Ochotonidae (Pikas)
172   Family Leporidae (Hares and Rabbits)

173   ORDER MACROSCELIDEA (Elephant Shrews)

173   Family Macroscelididae

175   Literature Cited
183   Index to Scientific Names
193   Index to Common Names

COMMON NAMES OF
MAMMALS OF THE WORLD

INTRODUCTION

HISTORY AND RATIONALE

The purpose of this book is to provide unique English common names for the mammals of the world. This volume contains English common names for all 4,629 mammal species recognized by Wilson and Reeder (1993) and, when appropriate, for genera, families, and orders. The lack of common names in that volume has proven frustrating to some users. This book is a response to those concerns. Although subspecies have importance in the study of evolutionary patterns, we made no attempt to assign a common name to any subspecies.

Of course, there are mammalogists who are averse to standardized common names, and there will be continuing debates about which names are appropriate for some of the more multi-nominal species. We encourage these debates and, to the extent that this book has fostered discussion, are content to see that we have generated continued interest in the topic. Nevertheless, many, if not most, mammalogists want a standard English reference. This book is an attempt to satisfy that reasonable desire. Additionally, we believe that a book of vernacular names may generate interest in mammals among a broader segment of the public. At least initially, scientific names may be confusing or unnecessary, and even impose barriers to beginning naturalists. A wider audience for mammalian biology may lead to increased interest in conservation and the preservation of critical habitats.

Currently, the American Ornithologists' Union checklist provides a standard reference for both scientific and common names of birds, and the list of guides to English common names includes such taxa as reptiles, amphibians, butterflies, vascular plants, aquatic invertebrates and fishes (e.g., AOU 1983; Cairns et al. 1991; Miller 1992; Monroe and Sibley 1993; Howard and Moore 1994). As the availability of standard lists of common names of organisms continues to increase, the need for a similar reference for mammals becomes apparent.

Some species with wide geographical distributions may have different common names in different parts of their range. Confusion may also arise when the same common name is applied to different species. Additionally, private and public agencies include common names as well as scientific names in their publications to increase their usefulness and to appeal to a wide audience of the scientifically curious public (Miller 1992). In general, common names may be more readily adaptable to lay uses than scientific names. Also, a readily available book of vernacular names for mammals would help restrict the proliferation of new names that may be unnecessary and confusing.

As with other books of common names for representatives of major taxonomic groups, we hope that this book will help to stabilize the use of vernacular names for mammal species and increase awareness of the diversity and geographical distribution of the world's mammal species. Many mammal species are rare, secretive in their behavior, located in remote areas, and not well studied. It will be appropriate in the future, as our knowledge of mammalian natural history expands, to revise selected common names to improve accuracy of species descriptions.

PRINCIPLES USED IN SELECTING COMMON NAMES

As a first step in compiling a book of unique English common names for all mammal species, we reviewed common names used in widely available general works on mammals looking for precedented usage (e.g., Macdonald 1984; Nowak 1999) or in regional publications (e.g., Nearctic—Banfield 1974; Hall 1981; Chapman and Feldhammer 1982; Ramirez-Pulido et al. 1982; Gonzalez and Leal 1984; Wilson and Ruff 1999; Neotropical—Goodwin 1946; Handley 1966; Hershkovitz 1972; Wilson 1983; Eisenberg 1989; Redford and Eisenberg 1992; Emmons 1997; Palearctic—Corbet 1978; Corbet and Ovenden 1980; Gromov and Baranova 1981; Corbet 1984; Niethammer and Krapp 1990; Ethiopian—Kingdon 1971–82, 1997; Smithers 1983; Oriental—Lekagul and McNeely 1977; Phillips 1980–84; Medway 1983; Heaney et al. 1987, 1998; Musser 1987; Corbet and Hill 1992; Roberts 1997; Australian—Ziegler 1982; Strahan 1995). We used Corbet and Hill 1991 and Nowak 1999 extensively as sources of names worldwide. Many of these common names were adopted to maintain consistency with previously published literature. In cases where several common names were used repeatedly, we generally selected the most frequently cited name, rather than the first to be applied, as the primary vernacular name for inclusion in our book. We also considered how well this name fit the principles described below. Approximately 65 percent of our names were obtained in this manner. Occasionally, we ignored a common name available in the literature for what we felt was a more apt name.

Most taxa in well-studied geographical areas (e.g., United States) already possess common names. Also, vernacular names are available for most large mammals (e.g., Proboscidea, Artiodactyla, Perissodactyla, Primates). Typically, our greatest challenge has been finding appropriate names for small- to medium-sized mammals living in the tropical areas of the world where few mammalian studies have been conducted.

If we could not identify a common name from our literature search, we created one for the species in question. We believe that common names should be informative about some distinctive aspect of the mammal's appearance, ecological habits, geographical distribution, or some combination of these aspects that is not cumbersome. In the process of generating new common names, we attempted to follow general guidelines similar to those used by Cairns et al. (1991). These guidelines are described below:

1. We attempted to maintain simplicity when possible and tried to avoid the use of hyphens and suffixes except when necessary.
2. We tried to avoid intimately linking the common name to the scientific name for a particular species, although we sometimes opted for consistency in including the generic name in some species names.
3. Modifiers that describe age or size (e.g., big, small, little, or large) that might have ambiguous meanings were avoided when possible and not already part of a commonly used vernacular name. Additionally, the expression "common" was avoided unless part of a well-established vernacular name.
4. Names honoring persons were avoided when possible because they lack descriptive value. However, when the scientific name is a patronym, we used it, written with an apostrophe.
5. We capitalized common names.
6. We tried to employ descriptive terms characteristic of that particular species. For example, structural attributes such as color or pattern are desirable descriptors and were frequently used in forming common names. Ecological characteristics were useful descriptors included in names. Also, many names include modifiers that refer to the geographical distribution of the species.
7. Generic names were employed outright in some cases.
8. Vernacular names derived from a language other than English were adopted when already well established.

9. Colorful, romantic, fanciful, metaphorical, and otherwise distinctive and original names were chosen as suggested by Cairns et al. (1991).

10. The duplication of common names between different major groups of animals (i.e., invertebrates and vertebrates) should be avoided (Cairns et al. 1991) and we did attempt to adhere to this principle to some extent. However, we did not conduct a complete survey for overlap.

11. Genera are given common names only when they contain more than one species. Otherwise the genus takes the common name of its only contained species. The same is true for families and orders.

GEOGRAPHIC DISTRIBUTION

For the purposes of this book we have divided the mammal-inhabited, large land masses of the world into six regions: *Nearctic* and *Neotropical* in the New World, and *Palearctic, Ethiopian, Oriental,* and *Australian* in the Old World (Cole et al. 1994). These regions correspond roughly to the zoogeographical realms (Darlington 1957), with a few changes noted below. We have placed the world's oceans and isolated islands into a seventh category, the *Oceanic* region.

We include the land from the Canadian Arctic south to the Mexican border with Guatemala in the *Nearctic* region, which is roughly equivalent to the Nearctic realm. The *Neotropical* region encompasses the mainland of Central and South America and the islands of the West Indies. These two regions comprise the New World.

We have partitioned the Old World into four regions: the Palearctic, Ethiopian, Oriental, and Australian. The *Palearctic* region encompasses the British Isles and the area bounded by the Atlantic Ocean in the west, the Arctic Ocean in the north, Mediterranean Africa, the Arabian Peninsula, and the Middle East in the south, and the Pacific Ocean in the east. The *Ethiopian* region includes sub-Saharan Africa and the island of Madagascar. The Indomalayan region and the Philippines south to Wallace's line comprise the *Oriental* region (Darlington 1957). The *Australian* region includes New Guinea, Australia, Tasmania, and adjacent islands. New Zealand and its islands were also included in this region.

The *Oceanic* region includes the isolated islands of the world, including the islands in the Arctic and Antarctic regions. Species inhabiting continental islands were included in the fauna for that continental region. The marine environment includes the coastal New and Old Worlds and the deep water areas of the Atlantic, Pacific, and Indian Oceans and the Arctic and Antarctic regions. The Antarctic region includes islands and oceans within the Antarctic convergence, and the Arctic region includes islands and oceans within the Arctic circle (including all of Greenland and Iceland).

Abbreviations for geographical regions are as follows (see map):

Nearctic Region = Nea   Oriental Region = Or
Neotropical Region = Neo   Australian Region = A
Palearctic Region = P   Oceanic Region = Oc
Ethiopian Region = E

HOW TO USE THIS BOOK

The taxonomic treatment and geographical distribution of each mammal species follow Wilson and Reeder (1993). Each of the 4,629 mammals species recognized by Wilson and Reeder (1993) was given a unique English common name. The literature cited section includes references consulted for common names as well as reference citations made in the text.

The book is presented in two columns, with the scientific classification for each species provided on the left (presented in phyletic order) and the region of the world where the species is found is given in parentheses; the second column contains the proposed vernacular name for the species. The second column is followed by a number in brackets that indicates a source in which the vernacular name was found during our literature search. These numbers correspond to the numbered references found in the Literature Cited. A blank in this column indicates a name that we generated.

ACKNOWLEDGMENTS

The taxonomy and geographical distributions used in this book of common names follow the second edition of *Mammal Species of the World,* a joint publication of the American Society of Mammalogists and the Smithsonian Institution (Wilson and Reeder 1993). We wish to thank the individual authors of that work: C. G. Anderson, R. L. Brownell, Jr., M. D. Carleton, F. Dieterlen, A. L. Gardner, C. P. Groves, P. Grubb, L. R. Heaney, R. S. Hoffmann, M. E. Holden, R. Hutterer, K. F. Koopman, J. G. Mead, G. G. Musser, J. L. Patton, D. A. Schlitter, R. W. Thorington, Jr., C. A. Woods, and W. C. Wozencraft. We are also grateful to those members of the Checklist Committee of the American Society of Mammalogists who reviewed draft lists of names and provided helpful suggestions (A. L. Gardner, C. P. Groves, P. Grubb, L. R. Heaney, J. G. Mead,

B. Patterson, J. L. Patton). R. S. Hoffmann kindly read the entire manuscript and provided us with numerous helpful suggestions. J. Payne, A. O'Connell, and M. Sanderson provided reliable assistance in tabulating and analyzing data. Daniel Cole and George Venable provided excellent cartographic assistance. This project was funded by the Office of Biodiversity Programs, Smithsonian Institution, and by grants to FRC from the Howard Hughes Medical Institute Fund, the Oak Fund, and the Division of Natural Sciences Research Fund, Colby College.

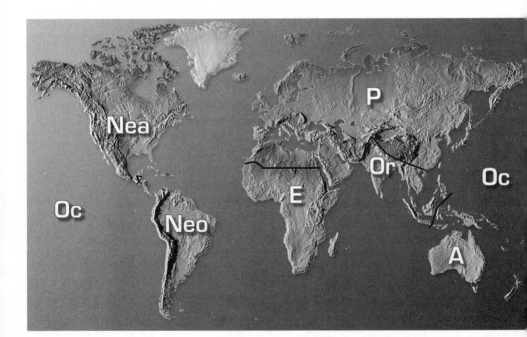

Regions of the Earth as referenced in the list of mammal common names. Nea = Nearctic, Oc = Oceanic, Neo = Neotropical, E = Ethiopian, P = Palearctic, Or = Oriental, A = Australian.

COMMON NAMES OF MAMMALS
OF THE WORLD

———————

ORDER MONOTREMATA   Monotremes   [55]
Family Tachyglossidae   Echidnas   [86]
   Genus *Tachyglossus*
     T. aculeatus (A)   Short-beaked Echidna   [55]
   Genus *Zaglossus*
     Z. bruijni (A)   Long-beaked Echidna   [55]
Family Ornithorhynchidae
   Genus *Ornithorhynchus*
     O. anatinus (A)   Platypus   [55]

ORDER DIDELPHIMORPHIA   American Opossums   [86]
Family Didelphidae
   Subfamily Caluromyinae
     Genus *Caluromys*   Woolly Opossums   [86]
      C. derbianus (Nea, Neo)   Central American Woolly
                                            Opossum   [26]
      C. lanatus (Neo)   Western Woolly Opossum   [26]
      C. philander (Neo)   Bare-tailed Woolly Opossum [26]
     Genus *Caluromysiops*
      C. irrupta (Neo)   Black-shouldered Opossum [26]
     Genus *Glironia*
      G. venusta (Neo)   Bushy-tailed Opossum   [26]
   Subfamily Didelphinae
     Genus *Chironectes*
      C. minimus (Nea, Neo)   Water Opossum   [26]
     Genus *Didelphis*   Large American Opossums   [5]
      D. albiventris (Neo)   White-eared Opossum   [26]
      D. aurita (Neo)   Big-eared Opossum

D. marsupialis (Nea, Neo)	Southern Opossum	[38]
D. virginiana (Nea, Neo)	Virginia Opossum	[114]
Genus *Gracilinanus*	Gracile Mouse Opossums	[26]
G. aceramarcae (Neo)	Aceramarca Gracile Mouse Opossum	
G. agilis (Neo)	Agile Gracile Mouse Opossum	[26]
G. dryas (Neo)	Wood Sprite Gracile Mouse Opossum	[26]
G. emiliae (Neo)	Emilia's Gracile Mouse Opossum	[26]
G. marica (Neo)	Northern Gracile Mouse Opossum	[26]
G. microtarsus (Neo)	Brazilian Gracile Mouse Opossum	[26]
Genus *Lestodelphys*		
L. halli (Neo)	Patagonian Opossum	[86]
Genus *Lutreolina*		
L. crassicaudata (Neo)	Lutrine Opossum	[26]
Genus *Marmosa*	Mouse Opossums	[86]
M. andersoni (Neo)	Anderson's Mouse Opossum	
M. canescens (Nea)	Grayish Mouse Opossum	[38]
M. lepida (Neo)	Little Rufous Mouse Opossum	[26]
M. mexicana (Nea, Neo)	Mexican Mouse Opossum	[98]
M. murina (Neo)	Murine Mouse Opossum	[26]
M. robinsoni (Neo)	Robinson's Mouse Opossum	[26]
M. rubra (Neo)	Red Mouse Opossum	[26]
M. tyleriana (Neo)	Tyler's Mouse Opossum	
M. xerophila (Neo)	Dryland Mouse Opossum	
Genus *Marmosops*	Slender Mouse Opossums	
M. cracens (Neo)	Slim-faced Slender Mouse Opossum	
M. dorothea (Neo)	Dorothy's Slender Mouse Opossum	[26]
M. fuscatus (Neo)	Gray-bellied Slender Mouse Opossum	[26]
M. handleyi (Neo)	Handley's Slender Mouse Opossum	

M. impavidus (Neo)	Andean Slender Mouse Opossum	[26]
M. incanus (Neo)	Gray Slender Mouse Opossum	[26]
M. invictus (Neo)	Slaty Slender Mouse Opossum	[26]
M. noctivagus (Neo)	White-bellied Slender Mouse Opossum	[26]
M. parvidens (Neo)	Delicate Slender Mouse Opossum	[26]
Genus Metachirus		
M. nudicaudatus (Nea, Neo)	Brown Four-eyed Opossum	[26]
Genus Micoureus	Woolly Mouse Opossums	[26]
M. alstoni (Neo)	Alston's Woolly Mouse Opossum	[26]
M. constantiae (Neo)	Pale-bellied Woolly Mouse Opossum	[26]
M. demerarae (Neo)	Long-furred Woolly Mouse Opossum	[26]
M. regina (Neo)	Short-furred Woolly Mouse Opossum	[26]
Genus Monodelphis	Short-tailed Opossums	[26]
M. adusta (Neo)	Sepia Short-tailed Opossum	[26]
M. americana (Neo)	Three-striped Short-tailed Opossum	[26]
M. brevicaudata (Neo)	Red-legged Short-tailed Opossum	[26]
M. dimidiata (Neo)	Southern Short-tailed Opossum	[26]
M. domestica (Neo)	Gray Short-tailed Opossum	[26]
M. emiliae (Neo)	Emilia's Short-tailed Opossum	[26]
M. iheringi (Neo)	Ihering's Short-tailed Opossum	[26]
M. kunsi (Neo)	Pygmy Short-tailed Opossum	[26]
M. maraxina (Neo)	Marajó Short-tailed Opossum	[26]
M. osgoodi (Neo)	Osgood's Short-tailed Opossum	[26]

M. rubida (Neo)	Chestnut-striped Short-tailed Opossum	[26]
M. scalops (Neo)	Long-nosed Short-tailed Opossum	[26]
M. sorex (Neo)	Shrewish Short-tailed Opossum	[26]
M. theresa (Neo)	Theresa's Short-tailed Opossum	[26]
M. unistriata (Neo)	One-striped Short-tailed Opossum	[26]
Genus *Philander*	Gray and Black Four-eyed Opossums	[86]
P. andersoni (Neo)	Black Four-eyed Opossum	[86]
P. opossum (Nea, Neo)	Gray Four-eyed Opossum	[86]
Genus *Thylamys*	Fat-tailed Opossums	
T. elegans (Neo)	Elegant Fat-tailed Opossum	
T. macrura (Neo)	Long-tailed Fat-tailed Opossum	
T. pallidior (Neo)	Pallid Fat-tailed Opossum	
T. pusilla (Neo)	Small Fat-tailed Opossum	
T. velutinus (Neo)	Velvety Fat-tailed Opossum	

ORDER PAUCITUBERCULATA Shrew Opossums
Family Caenolestidae

Genus *Caenolestes*	Northern Shrew Opossums	[109]
C. caniventer (Neo)	Gray-bellied Shrew Opossum	
C. convelatus (Neo)	Blackish Shrew Opossum	
C. fuliginosus (Neo)	Silky Shrew Opossum	
Genus *Lestoros*		
L. inca (Neo)	Incan Shrew Opossum	
Genus *Rhyncholestes*		
R. raphanurus (Neo)	Chilean Shrew Opossum	[86]

ORDER MICROBIOTHERIA
Family Microbiotheriidae

Genus *Dromiciops*		
D. gliroides (Neo)	Monito Del Monte	[86]

ORDER DASYUROMORPHIA   Carnivorous Marsupials

Family Thylacinidae
    Genus *Thylacinus*
        T. cynocephalus (A)   Thylacine   [55]
Family Myrmecobiidae
    Genus *Myrmecobius*
        M. fasciatus (A)   Numbat   [110]
Family Dasyuridae   Dasyurids   [55]
    Genus *Antechinus*   Antechinuses   [109]
        A. bellus (A)   Fawn Antechinus   [110]
        A. flavipes (A)   Yellow-footed
                 Antechinus   [110]
        A. godmani (A)   Atherton Antechinus   [110]
        A. leo (A)   Cinnamon Antechinus   [110]
        A. melanurus (A)   Black-tailed Antechinus   [55]
        A. minimus (A)   Swamp Antechinus   [110]
        A. naso (A)   Long-nosed Antechinus   [55]
        A. stuartii (A)   Brown Antechinus   [110]
        A. swainsonii (A)   Dusky Antechinus   [110]
        A. wilhelmina (A)   Lesser Antechinus   [16]
    Genus *Dasycercus*   Crested-tailed Marsupial
                 Mice   [109]
        D. byrnei (A)   Kowari   [55]
        D. cristicauda (A)   Mulgara   [110]
    Genus *Dasykaluta*
        D. rosamondae (A)   Little Red Kaluta   [110]
    Genus *Dasyurus*   Quolls   [16]
        D. albopunctatus (A)   New Guinean Quoll   [55]
        D. geoffroii (A)   Western Quoll   [110]
        D. hallucatus (A)   Northern Quoll   [110]
        D. maculatus (A)   Spotted-tailed Quoll   [110]
        D. spartacus (A)   Bronze Quoll   [55]
        D. viverrinus (A)   Eastern Quoll   [110]
    Genus *Murexia*   Long-tailed Dasyures   [109]
        M. longicaudata (A)   Short-furred Dasyure   [28]
        M. rothschildi (A)   Broad-striped Dasyure   [28]
    Genus *Myoictis*
        M. melas (A)   Three-striped Dasyure   [28]
    Genus *Neophascogale*
        N. lorentzi (A)   Speckled Dasyure   [28]

Genus *Ningaui*	Ningauis	
N. *ridei* (A)	Wongai Ningaui	[110]
N. *timealeyi* (A)	Pilbara Ningaui	[110]
N. *yvonnae* (A)	Southern Ningaui	[110]
Genus *Parantechinus*	Dibblers	
P. *apicalis* (A)	Southern Dibbler	[110]
P. *bilarni* (A)	Sandstone Dibbler	[55]
Genus *Phascogale*	Phascogales	[109]
P. *calura* (A)	Red-tailed Phascogale	[110]
P. *tapoatafa* (A)	Brush-tailed Phascogale	[110]
Genus *Phascolosorex*	Marsupial Shrews	
P. *doriae* (A)	Red-bellied Marsupial Shrew	
P. *dorsalis* (A)	Narrow-striped Marsupial Shrew	
Genus *Planigale*	Planigales	[16]
P. *gilesi* (A)	Paucident Planigale	[110]
P. *ingrami* (A)	Long-tailed Planigale	[110]
P. *maculata* (A)	Pygmy Planigale	[16]
P. *novaeguineae* (A)	New Guinean Planigale	[55]
P. *tenuirostris* (A)	Narrow-nosed Planigale	[110]
Genus *Pseudantechinus*	Pseudantechinuses	
P. *macdonnellensis* (A)	Fat-tailed Pseudantechinus	[110]
P. *ningbing* (A)	Ningbing Pseudantechinus	[110]
P. *woolleyae* (A)	Woolley's Pseudantechinus	[110]
Genus *Sarcophilus*		
S. *laniarius* (A)	Tasmanian Devil	[110]
Genus *Sminthopsis*	Dunnarts	[16]
S. *aitkeni* (A)	Kangaroo Island Dunnart	[110]
S. *archeri* (A)	Chestnut Dunnart	[110]
S. *butleri* (A)	Carpentarian Dunnart	[55]
S. *crassicaudata* (A)	Fat-tailed Dunnart	[110]
S. *dolichura* (A)	Little Long-tailed Dunnart	[110]
S. *douglasi* (A)	Julia Creek Dunnart	[110]
S. *fuliginosus* (A)	Sooty Dunnart	
S. *gilberti* (A)	Gilbert's Dunnart	[110]
S. *granulipes* (A)	White-tailed Dunnart	[110]
S. *griseoventer* (A)	Gray-bellied Dunnart	[110]
S. *hirtipes* (A)	Hairy-footed Dunnart	[110]
S. *laniger* (A)	Kultarr	[55]

S. leucopus (A)	White-footed Dunnart	[110]
S. longicaudata (A)	Long-tailed Dunnart	[110]
S. macroura (A)	Stripe-faced Dunnart	[110]
S. murina (A)	Slender-tailed Dunnart	
S. ooldea (A)	Ooldea Dunnart	[110]
S. psammophila (A)	Sandhill Dunnart	[110]
S. virginiae (A)	Red-cheeked Dunnart	[110]
S. youngsoni (A)	Lesser Hairy-footed Dunnart	[110]

ORDER PERAMELEMORPHIA	Bandicoots and Bilbies	[110]
Family Peramelidae	Australian Bandicoots and Bilbies	[55]
Genus *Chaeropus*		
C. ecaudatus (A)	Pig-footed Bandicoot	[110]
Genus *Isoodon*	Short-nosed Bandicoots	[86]
I. auratus (A)	Golden Bandicoot	[110]
I. macrourus (A)	Northern Brown Bandicoot	[110]
I. obesulus (A)	Southern Brown Bandicoot	[110]
Genus *Macrotis*	Bilbies	[86]
M. lagotis (A)	Bilby	[110]
M. leucura (A)	Lesser Bilby	[110]
Genus *Perameles*	Long-nosed Bandicoots	[86]
P. bougainville (A)	Western Barred Bandicoot	[110]
P. eremiana (A)	Desert Bandicoot	[110]
P. gunnii (A)	Eastern Barred Bandicoot	[110]
P. nasuta (A)	Long-nosed Bandicoot	[110]
Family Peroryctidae	Rainforest Bandicoots	[55]
Genus *Echymipera*	Echymiperas	
E. clara (A)	Clara's Echymipera	[28]
E. davidi (A)	David's Echymipera	[29]
E. echinista (A)	Menzies's Echymipera	[28]
E. kalubu (A)	Kalubu Echymipera	
E. rufescens (A)	Long-nosed Echymipera	[28]
Genus *Microperoryctes*	Mouse Bandicoots	[16]
M. longicauda (A)	Striped Bandicoot	[28]
M. murina (A)	Mouse Bandicoot	[28]
M. papuensis (A)	Papuan Bandicoot	[28]

Genus *Peroryctes*	New Guinean Bandicoots	[16]
P. broadbenti (A)	Giant Bandicoot	[55]
P. raffrayana (A)	Raffray's Bandicoot	[28]
Genus *Rhynchomeles*		
R. prattorum (A)	Ceram Bandicoot	[55]

ORDER NOTORYCTEMORPHIA

ORDER NOTORYCTEMORPHIA	Marsupial Moles	[55]
Family Notoryctidae		
Genus *Notoryctes*		
N. caurinus (A)	Northwestern Marsupial Mole	
N. typhlops (A)	Marsupial Mole	[110]

ORDER DIPROTODONTIA

ORDER DIPROTODONTIA	Koalas, Wombats, Possums, Macropods, and relatives	
Family Phascolarctidae		
Genus *Phascolarctos*		
P. cinereus (A)	Koala	[110]
Family Vombatidae	Wombats	[55]
Genus *Lasiorhinus*	Hairy-nosed Wombats	[86]
L. krefftii (A)	Northern Hairy-nosed Wombat	[110]
L. latifrons (A)	Southern Hairy-nosed Wombat	[110]
Genus *Vombatus*		
V. ursinus (A)	Coarse-haired Wombat	[86]
Family Phalangeridae	Cuscuses and Brushtail Possums	[55]
Genus *Ailurops*		
A. ursinus (Or, A)	Bear Cuscus	
Genus *Phalanger*	Cuscuses	[16]
P. carmelitae (A)	Mountain Cuscus	[55]
P. lullulae (A)	Woodlark Cuscus	[16]
P. matanim (A)	Telefomin Cuscus	[55]
P. orientalis (Or, A)	Gray Cuscus	
P. ornatus (A)	Moluccan Cuscus	[55]
P. pelengensis (Or, A)	Peleng Island Cuscus	[55]
P. rothschildi (A)	Obi Island Cuscus	[55]
P. sericeus (A)	Silky Cuscus	[55]
P. vestitus (A)	Stein's Cuscus	[55]

Genus *Spilocuscus*	Spotted Cuscuses	[86]
S. *maculatus* (A)	Short-tailed Spotted Cuscus	
S. *rufoniger* (A)	Black-spotted Cuscus	[55]
Genus *Strigocuscus*	Plain Cuscuses	
S. *celebensis* (A)	Little Celebes Cuscus	[55]
S. *gymnotis* (A)	Ground Cuscus	[28]
Genus *Trichosurus*	Brushtail Possums	
T. *arnhemensis* (A)	Northern Brushtail Possum	[110]
T. *caninus* (A)	Mountain Brushtail Possum	[110]
T. *vulpecula* (A)	Silver-gray Brushtail Possum	
Genus *Wyulda*		
W. *squamicaudata* (A)	Scaly-tailed Possum	[110]
Family Potoroidae	Bettongs, Rat Kangaroos, and Potoroos	
Genus *Aepyprymnus*		
A. *rufescens* (A)	Rufous Bettong	[110]
Genus *Bettongia*	Bettongs	[16]
B. *gaimardi* (A)	Tasmanian Bettong	[110]
B. *lesueur* (A)	Burrowing Bettong	[110]
B. *penicillata* (A)	Brush-tailed Bettong	[110]
Genus *Caloprymnus*		
C. *campestris* (A)	Desert Rat Kangaroo	[110]
Genus *Hypsiprymnodon*		
H. *moschatus* (A)	Musky Rat Kangaroo	[110]
Genus *Potorous*	Potoroos	[16]
P. *longipes* (A)	Long-footed Potoroo	[110]
P. *platyops* (A)	Broad-faced Potoroo	[110]
P. *tridactylus* (A)	Long-nosed Potoroo	[110]
Family Macropodidae	Kangaroos and Wallabies	[55]
Genus *Dendrolagus*	Tree Kangaroos	[16]
D. *bennettianus* (A)	Bennett's Tree Kangaroo	[110]
D. *dorianus* (A)	Doria's Tree Kangaroo	[55]
D. *goodfellowi* (A)	Goodfellow's Tree Kangaroo	[55]
D. *inustus* (A)	Grizzled Tree Kangaroo	[55]
D. *lumholtzi* (A)	Lumholtz's Tree Kangaroo	[110]
D. *matschiei* (A)	Huon Tree Kangaroo	[28]
D. *scottae* (A)	Tenkile Tree Kangaroo	[28]

D. spadix (A)	Lowland Tree Kangaroo	[55]
D. ursinus (A)	White-throated Tree Kangaroo	[55]
Genus *Dorcopsis*	Dorcopsises	
D. atrata (A)	Black Dorcopsis	[28]
D. hageni (A)	White-striped Dorcopsis	[28]
D. luctuosa (A)	Gray Dorcopsis	[28
D. muelleri (A)	Brown Dorcopsis	[28]
Genus *Dorcopsulus*	Forest Wallabies	
D. macleayi (A)	Papuan Forest Wallaby	[16]
D. vanheurni (A)	Lesser Forest Wallaby	[16]
Genus *Lagorchestes*	Hare-wallabies	[16]
L. asomatus (A)	Central Hare-wallaby	[110]
L. conspicillatus (A)	Spectacled Hare-wallaby	[110]
L. hirsutus (A)	Rufous Hare-wallaby	[110]
L. leporides (A)	Eastern Hare-wallaby	[110]
Genus *Lagostrophus*		
L. fasciatus (A)	Banded Hare-wallaby	[110]
Genus *Macropus*	Kangaroos, Wallaroos and Wallabies	
M. agilis (A)	Agile Wallaby	[110]
M. antilopinus (A)	Antilopine Wallaroo	[110]
M. bernardus (A)	Black Wallaroo	[110]
M. dorsalis (A)	Black-striped Wallaby	[55]
M. eugenii (A)	Tammar Wallaby	[110]
M. fuliginosus (A)	Western Gray Kangaroo	[110]
M. giganteus (A)	Eastern Gray Kangaroo	[110]
M. greyi (A)	Toolache Wallaby	[110]
M. irma (A)	Western Brush Wallaby	[110]
M. parma (A)	Parma Wallaby	[110]
M. parryi (A)	Whiptail Wallaby	[110]
M. robustus (A)	Hill Wallaroo	
M. rufogriseus (A)	Red-necked Wallaby	[110]
M. rufus (A)	Red Kangaroo	[110]
Genus *Onychogalea*	Nail-tailed Wallabies	[16]
O. fraenata (A)	Bridled Nail-tailed Wallaby	[110]
O. lunata (A)	Crescent Nail-tailed Wallaby	[110]
O. unguifera (A)	Northern Nail-tailed Wallaby	[110]

Genus *Petrogale* | Rock Wallabies | [16]
P. *assimilis* (A) | Allied Rock Wallaby | [110]
P. *brachyotis* (A) | Short-eared Rock Wallaby | [110]
P. *burbidgei* (A) | Monjon | [110]
P. *concinna* (A) | Pygmy Rock Wallaby | [109]
P. *godmani* (A) | Godman's Rock Wallaby | [110]
P. *inornata* (A) | Unadorned Rock Wallaby | [110]
P. *lateralis* (A) | Black-footed Rock Wallaby | [110]
P. *penicillata* (A) | Brush-tailed Rock Wallaby | [110]
P. *persephone* (A) | Proserpine Rock Wallaby | [110]
P. *rothschildi* (A) | Rothschild's Rock Wallaby | [110]
P. *xanthopus* (A) | Yellow-footed Rock Wallaby | [110]

Genus *Setonix*
S. *brachyurus* (A) | Quokka | [110]
Genus *Thylogale* | Pademelons | [16]
T. *billardierii* (A) | Tasmanian Pademelon | [110]
T. *brunii* (A) | Dusky Pademelon | [55]
T. *stigmatica* (A) | Red-legged Pademelon | [110]
T. *thetis* (A) | Red-necked Pademelon | [110]
Genus *Wallabia*
W. *bicolor* (A) | Swamp Wallaby | [110]
Family Burramyidae | Pygmy Possums | [55]
Genus *Burramys*
B. *parvus* (A) | Mountain Pygmy Possum | [110]
Genus *Cercartetus* | Pygmy Possums | [110]
C. *caudatus* (A) | Long-tailed Pygmy Possum | [110]
C. *concinnus* (A) | Western Pygmy Possum | [110]
C. *lepidus* (A) | Tasmanian Pygmy Possum | [16]
C. *nanus* (A) | Eastern Pygmy Possum | [110]
Family Pseudocheiridae | Ringtail Possums | [55]
Genus *Hemibelideus*
H. *lemuroides* (A) | Lemuroid Ringtail Possum | [110]
Genus *Petauroides*
P. *volans* (A) | Greater Glider | [110]
Genus *Petropseudes*
P. *dahli* (A) | Rock Ringtail Possum | [110]
Genus *Pseudocheirus* | Ringtails | [16]
P. *canescens* (A) | Daintree River Ringtail | [110]
P. *caroli* (A) | Weyland Ringtail | [55]

P. forbesi (A)	Moss-forest Ringtail	[16]
P. herbertensis (A)	Herbert River Ringtail	[16]
P. mayeri (A)	Pygmy Ringtail	[55]
P. peregrinus (A)	Queensland Ringtail	[109]
P. schlegeli (A)	Arfak Ringtail	[55]
Genus *Pseudochirops*	Ringtail Possums	[110]
P. albertisii (A)	d'Albertis's Ringtail Possum	
P. archeri (A)	Green Ringtail Possum	[110]
P. corinnae (A)	Golden Ringtail Possum	
P. cupreus (A)	Coppery Ringtail Possum	
Family Petauridae	Gliding and Striped Possums	[86]
Genus *Dactylopsila*	Striped Possums	[16]
D. megalura (A)	Great-tailed Triok	[28]
D. palpator (A)	Long-fingered Triok	[28]
D. tatei (A)	Tate's Triok	[29]
D. trivirgata (A)	Striped Possum	[110]
Genus *Gymnobelideus*		
G. leadbeateri (A)	Leadbeater's Possum	[110]
Genus *Petaurus*	Lesser Gliding Possums	[86]
P. abidi (A)	Northern Glider	[55]
P. australis (A)	Yellow-bellied Glider	[110]
P. breviceps (A)	Sugar Glider	[110]
P. gracilis (A)	Mahogany Glider	[110]
P. norfolcensis (A)	Squirrel Glider	[110]
Family Tarsipedidae		
Genus *Tarsipes*		
T. rostratus (A)	Honey Possum	[110]
Family Acrobatidae	Feathertail Gliders and Possums	
Genus *Acrobates*		
A. pygmaeus (A)	Feathertail Glider	[110]
Genus *Distoechurus*		
D. pennatus (A)	Feathertail Possum	[55]
ORDER XENARTHRA	Edentates	[16]
Family Bradypodidae	Three-toed Sloths	[24]
Genus *Bradypus*		
B. torquatus (Neo)	Maned Three-toed Sloth	[26]

B. tridactylus (Neo)	Pale-throated Three-toed Sloth	[26]
B. variegatus (Neo)	Brown-throated Three-toed Sloth	[26]
Family Megalonychidae	Two-toed Sloths	[26]
Genus Choloepus		
C. didactylus (Neo)	Southern Two-toed Sloth	[26]
C. hoffmanni (Neo)	Hoffmann's Two-toed Sloth	[26]
Family Dasypodidae	Armadillos	[26]
Subfamily Chlamyphorinae		
Genus Chlamyphorus	Fairy Armadillos	[16]
C. retusus (Neo)	Chacoan Fairy Armadillo	[16]
C. truncatus (Neo)	Pink Fairy Armadillo	[16]
Subfamily Dasypodinae		
Genus Cabassous	Naked-tailed Armadillos	[26]
C. centralis (Nea, Neo)	Northern Naked-tailed Armadillo	[26]
C. chacoensis (Neo)	Chacoan Naked-tailed Armadillo	[7]
C. tatouay (Neo)	Greater Naked-tailed Armadillo	[97]
C. unicinctus (Neo)	Southern Naked-tailed Armadillo	[26]
Genus Chaetophractus	Hairy Armadillos	[16]
C. nationi (Neo)	Andean Hairy Armadillo	[16]
C. vellerosus (Neo)	Screaming Hairy Armadillo	
C. villosus (Neo)	Large Hairy Armadillo	
Genus Dasypus	Long-nosed Armadillos	[16]
D. hybridus (Neo)	Southern Long-nosed Armadillo	[16]
D. kappleri (Neo)	Great Long-nosed Armadillo	[26]
D. novemcinctus (Nea, Neo)	Nine-banded Armadillo	[114]
D. pilosus (Neo)	Hairy Long-nosed Armadillo	[16]
D. sabanicola (Neo)	Llanos Long-nosed Armadillo	[16]
D. septemcinctus (Neo)	Seven-banded Armadillo	[26]
Genus Euphractus		
E. sexcinctus (Neo)	Six-banded Armadillo	[7]

Genus *Priodontes*

    P. maximus (Neo)   Giant Armadillo   [26]

Genus *Tolypeutes*   Three-banded Armadillos   [16]

    T. matacus (Neo)   Southern Three-banded
Armadillo   [16]

    T. tricinctus (Neo)   Brazilian Three-banded
Armadillo   [16]

Genus *Zaedyus*

    Z. pichiy (Neo)   Pichi   [16]

Family Myrmecophagidae   American Anteaters   [16]

Genus *Cyclopes*

    C. didactylus (Nea, Neo)   Silky Anteater   [98]

Genus *Myrmecophaga*

    M. tridactyla (Neo)   Giant Anteater   [26]

Genus *Tamandua*   Tamanduas   [26]

    T. mexicana (Nea, Neo)   Northern Tamandua   [26]

    T. tetradactyla (Neo)   Southern Tamandua   [26]

ORDER INSECTIVORA   Insectivores   [86]

Family Solenodontidae   Solenodons   [86]

Genus *Solenodon*

    S. cubanus (Neo)   Cuban Solenodon   [16]

    S. marcanoi (Neo)   Marcano's Solenodon

    S. paradoxus (Neo)   Hispaniolan Solenodon

Family Nesophontidae   Nesophontes

Genus *Nesophontes*

    N. edithae (Neo)   Puerto Rican Nesophontes

    N. hypomicrus (Neo)   Atalaye Nesophontes   [37]

    N. longirostris (Neo)   Slender Cuban Nesophontes

    N. major (Neo)   Greater Cuban
Nesophontes

    N. micrus (Neo)   Western Cuban
Nesophontes   [37]

    N. paramicrus (Neo)   St. Michel Nesophontes   [37]

    N. submicrus (Neo)   Lesser Cuban Nesophontes

    N. zamicrus (Neo)   Haitian Nesophontes   [37]

Family Tenrecidae   Tenrecs   [84]

Subfamily Geogalinae

Genus *Geogale*

    G. aurita (E)   Large-eared Tenrec   [84]

Subfamily Oryzorictinae
  Genus *Limnogale*
    L. mergulus (E)   Aquatic Tenrec   [84]
  Genus *Microgale*   Shrew Tenrecs   [16]
    M. brevicaudata (E)   Short-tailed Shrew Tenrec   [84]
    M. cowani (E)   Cowan's Shrew Tenrec   [84]
    M. dobsoni (E)   Dobson's Shrew Tenrec   [16]
    M. dryas (E)   Tree Shrew Tenrec
    M. gracilis (E)   Gracile Shrew Tenrec   [84]
    M. longicaudata (E)   Lesser Long-tailed Shrew
                                     Tenrec   [84]
    M. parvula (E)   Pygmy Shrew Tenrec   [84]
    M. principula (E)   Greater Long-tailed Shrew
                                     Tenrec   [84]
    M. pulla (E)   Dark Shrew Tenrec
    M. pusilla (E)   Least Shrew Tenrec
    M. talazaci (E)   Talazac's Shrew Tenrec   [16]
    M. thomasi (E)   Thomas's Shrew Tenrec   [84]
  Genus *Oryzorictes*   Rice Tenrecs   [16]
    O. hova (E)   Hova Rice Tenrec
    O. talpoides (E)   Molelike Rice Tenrec   [109]
    O. tetradactylus (E)   Four-toed Rice Tenrec
Subfamily Potamogalinae
  Genus *Micropotamogale*   Dwarf Otter Shrews   [86]
    M. lamottei (E)   Nimba Otter Shrew   [84]
    M. ruwenzorii (E)   Ruwenzori Otter Shrew   [84]
  Genus *Potamogale*
    P. velox (E)   Giant Otter Shrew   [16]
Subfamily Tenrecinae
  Genus *Echinops*
    E. telfairi (E)   Lesser Hedgehog Tenrec   [16]
  Genus *Hemicentetes*
    H. semispinosus (E)   Streaked Tenrec   [16]
  Genus *Setifer*
    S. setosus (E)   Greater Hedgehog Tenrec   [84]
  Genus *Tenrec*
    T. ecaudatus (E)   Tailless Tenrec   [16]
Family Chrysochloridae   Golden Moles   [84]
  Genus *Amblysomus*   South African Golden Moles   [86]
    A. gunningi (E)   Gunning's Golden Mole   [84]

A. hottentotus (E)	Hottentot Golden Mole	[16]
A. iris (E)	Zulu Golden Mole	[84]
A. julianae (E)	Juliana's Golden Mole	[84]
Genus *Calcochloris*		
C. obtusirostris (E)	Yellow Golden Mole	[16]
Genus *Chlorotalpa*	African Golden Moles	[86]
C. arendsi (E)	Arend's Golden Mole	[16]
C. duthieae (E)	Duthie's Golden Mole	[78]
C. leucorhina (E)	Congo Golden Mole	[16]
C. sclateri (E)	Sclater's Golden Mole	[16]
C. tytonis (E)	Somali Golden Mole	[16]
Genus *Chrysochloris*	Cape Golden Moles	[86]
C. asiatica (E)	Cape Golden Mole	[16]
C. stuhlmanni (E)	Stuhlmann's Golden Mole	[16]
C. visagiei (E)	Visagie's Golden Mole	[84]
Genus *Chrysospalax*	Large Golden Moles	[86]
C. trevelyani (E)	Giant Golden Mole	[84]
C. villosus (E)	Rough-haired Golden Mole	[84]
Genus *Cryptochloris*	Secretive Golden Moles	
C. wintoni (E)	De Winton's Golden Mole	[84]
C. zyli (E)	Van Zyl's Golden Mole	[84]
Genus *Eremitalpa*		
E. granti (E)	Grant's Golden Mole	[84]
Family Erinaceidae	Hedgehogs and Moonrats	[16]
Subfamily Erinaceinae		
Genus *Atelerix*	African Hedgehogs	[16]
A. albiventris (E)	Four-toed Hedgehog	[16]
A. algirus (P)	North African Hedgehog	
A. frontalis (E)	Southern African Hedgehog	[78]
A. sclateri (E)	Somali Hedgehog	[16]
Genus *Erinaceus*	Eurasian Hedgehogs	[86]
E. amurensis (P)	Amur Hedgehog	[17]
E. concolor (P)	Eastern European Hedgehog	[16]
E. europaeus (P)	Western European Hedgehog	[16]
Genus *Hemiechinus*	Desert Hedgehogs	
H. aethiopicus (P, E)	Desert Hedgehog	[109]
H. auritus (P)	Long-eared Hedgehog	[16]

H. collaris (P, Or)	Indian Long-eared Hedgehog	[16]
H. hypomelas (P)	Brandt's Hedgehog	[17]
H. micropus (P, Or)	Indian Hedgehog	[17]
H. nudiventris (Or)	Bare-bellied Hedgehog	
Genus *Mesechinus*	Steppe Hedgehogs	
M. dauuricus (P)	Daurian Hedgehog	[16]
M. hughi (P)	Hugh's Hedgehog	
Subfamily Hylomyinae		
Genus *Echinosorex*		
E. gymnura (Or)	Moonrat	[16]
Genus *Hylomys*	Asian Gymnures	
H. hainanensis (P)	Hainan Gymnure	
H. sinensis (P, Or)	Shrew Gymnure	
H. suillus (Or)	Short-tailed Gymnure	[109]
Genus *Podogymnura*	Philippine Gymnures	[86]
P. aureospinula (Or)	Dinagat Gymnure	[48]
P. truei (Or)	Mindanao Gymnure	[48]
Family Soricidae	Shrews	[86]
Subfamily Crocidurinae		
Genus *Congosorex*		
C. polli (E)	Poll's Shrew	
Genus *Crocidura*	White-toothed Shrews	[16]
C. aleksandrisi (P)	Alexandrian Shrew	
C. allex (E)	Highland Shrew	
C. andamanensis (Or)	Andaman Shrew	[109]
C. ansellorum (E)	Ansell's Shrew	
C. arabica (P)	Arabian Shrew	
C. armenica (P)	Armenian Shrew	
C. attenuata (P, Or)	Indochinese Shrew	[48]
C. attila (E)	Hun Shrew	
C. baileyi (E)	Bailey's Shrew	
C. batesi (E)	Bates's Shrew	
C. beatus (Or)	Mindanao Shrew	
C. beccarii (Or)	Beccari's Shrew	
C. bottegi (E)	Bottego's Shrew	[41]
C. bottegoides (E)	Bale Shrew	
C. buettikoferi (E)	Buettikofer's Shrew	[41]
C. caliginea (E)	African Foggy Shrew	
C. canariensis (P)	Canary Shrew	

C. cinderella (E)	Cinderella Shrew	
C. congobelgica (E)	Congo Shrew	
C. cossyrensis (P)	Pantellerian Shrew	
C. crenata (E)	Long-footed Shrew	
C. crossei (E)	Crosse's Shrew	[41]
C. cyanea (E)	Reddish-gray Musk Shrew	[78]
C. denti (E)	Dent's Shrew	
C. desperata (E)	Desperate Shrew	
C. dhofarensis (P)	Dhofarian Shrew	
C. dolichura (E)	Long-tailed Musk Shrew	[41]
C. douceti (E)	Doucet's Musk Shrew	
C. dsinezumi (P)	Dsinezumi Shrew	
C. eisentrauti (E)	Eisentraut's Shrew	
C. elgonius (E)	Elgon Shrew	
C. elongata (A)	Elongated Shrew	
C. erica (E)	Heather Shrew	
C. fischeri (E)	Fischer's Shrew	
C. flavescens (E)	Greater Red Musk Shrew	[78]
C. floweri (P)	Flower's Shrew	[16]
C. foxi (E)	Fox's Shrew	
C. fuliginosa (P, Or)	Southeast Asian Shrew	
C. fulvastra (E)	Savanna Shrew	[41]
C. fumosa (E)	Smoky White-toothed Shrew	
C. fuscomurina (E)	Tiny Musk Shrew	[78]
C. glassi (E)	Glass's Shrew	
C. goliath (E)	Goliath Shrew	
C. gracilipes (E)	Peters's Musk Shrew	[16]
C. grandiceps (E)	Large-headed Shrew	
C. grandis (Or)	Mt. Malindang Shrew	
C. grassei (E)	Grasse's Shrew	
C. grayi (Or)	Luzon Shrew	[48]
C. greenwoodi (E)	Greenwood's Shrew	
C. gueldenstaedtii (P)	Gueldenstaedt's Shrew	
C. harenna (E)	Harenna Shrew	
C. hildegardeae (E)	Hildegarde's Shrew	
C. hirta (E)	Lesser Red Musk Shrew	[78]
C. hispida (Or)	Andaman Spiny Shrew	[17]
C. horsfieldii (P, Or)	Horsfield's Shrew	[17]
C. jacksoni (E)	Jackson's Shrew	

C. jenkinsi (Or)	Jenkin's Shrew	
C. kivuana (E)	Kivu Shrew	
C. lamottei (E)	Lamotte's Shrew	[41]
C. lanosa (E)	Lemara Shrew	
C. lasiura (P)	Ussuri White-toothed Shrew	
C. latona (E)	Latona Shrew	
C. lea (A)	Sulawesi Shrew	
C. leucodon (P)	Bicolored Shrew	
C. levicula (A)	Celebes Shrew	
C. littoralis (E)	Butiaba Naked-tailed Shrew	[49]
C. longipes (E)	Savanna Swamp Shrew	[41]
C. lucina (E)	Moorland Shrew	
C. ludia (E)	Dramatic Shrew	
C. luna (E)	Greater Gray-brown Musk Shrew	[78]
C. lusitania (E)	Mauritanian Shrew	[41]
C. macarthuri (E)	MacArthur's Shrew	
C. macmillani (E)	MacMillan's Shrew	
C. macowi (E)	Macow's Shrew	
C. malayana (Or)	Malayan Shrew	
C. manengubae (E)	Manenguba Shrew	
C. maquassiensis (E)	Maquassie Musk Shrew	[78]
C. mariquensis (E)	Swamp Musk Shrew	[78]
C. maurisca (E)	Dark Shrew	
C. maxi (Or, A)	Max's Shrew	
C. mindorus (Or)	Mindoro Shrew	[48]
C. minuta (Or)	Minute Shrew	
C. miya (Or)	Sri Lankan Long-tailed Shrew	[17]
C. monax (E)	Rombo Shrew	
C. monticola (Or)	Sunda Shrew	[74]
C. montis (E)	Montane White-toothed Shrew	
C. muricauda (E)	Mouse-tailed Shrew	
C. mutesae (E)	Uganda Large-toothed Shrew	[49]
C. nana (E)	Dwarf White-toothed Shrew	
C. nanilla (E)	Tiny White-toothed Shrew	
C. neglecta (Or)	Neglected Shrew	

C. negrina (Or)	Negros Shrew	[48]
C. nicobarica (Or)	Nicobar Shrew	[17]
C. nigeriae (E)	Nigerian Shrew	
C. nigricans (E)	Black White-toothed Shrew	
C. nigripes (A)	Black-footed Shrew	
C. nigrofusca (E)	Tenebrous Shrew	
C. nimbae (E)	Nimba Shrew	
C. niobe (E)	Stony Shrew	
C. obscurior (E)	Obscure White-toothed Shrew	
C. olivieri (P, E)	Olivier's Shrew	
C. orii (P)	Amami Shrew	[17]
C. osorio (P)	Osorio Shrew	
C. palawanensis (Or)	Palawan Shrew	[48]
C. paradoxura (Or)	Paradox Shrew	
C. parvipes (E)	Small-footed Shrew	
C. pasha (E)	Pasha Shrew	
C. pergrisea (Or)	Pale Gray Shrew	
C. phaeura (E)	Guramba Shrew	
C. picea (E)	Pitch Shrew	
C. pitmani (E)	Pitman's Shrew	
C. planiceps (E)	Flat-headed Shrew	
C. poensis (E)	Fraser's Musk Shrew	[41]
C. polia (E)	Fuscous Shrew	
C. pullata (P, Or)	Dusky Shrew	
C. raineyi (E)	Rainey Shrew	[49]
C. religiosa (P)	Egyptian Pygmy Shrew	[16]
C. rhoditis (A)	Temboan Shrew	
C. roosevelti (E)	Roosevelt's Shrew	
C. russula (P)	White-toothed Shrew	[109]
C. selina (E)	Moon Shrew	
C. serezkyensis (P)	Serezkaya Shrew	
C. sibirica (P)	Siberian Shrew	
C. sicula (P)	Sicilian Shrew	
C. silacea (E)	Lesser Gray-brown Musk Shrew	[78]
C. smithii (E)	Desert Musk Shrew	[109]
C. somalica (E)	Somali Shrew	
C. stenocephala (E)	Narrow-headed Shrew	
C. suaveolens (P)	Lesser Shrew	

C. susiana (P)	Iranian Shrew	
C. tansaniana (E)	Tanzanian Shrew	
C. tarella (E)	Ugandan Shrew	
C. tarfayensis (P, E)	Tarfaya Shrew	
C. telfordi (E)	Telford's Shrew	
C. tenuis (A)	Thin Shrew	
C. thalia (E)	Thalia Shrew	
C. theresae (E)	Therese's Shrew	
C. thomensis (E)	Sao Tomé Shrew	
C. turba (E)	Tumultuous Shrew	
C. ultima (E)	Ultimate Shrew	
C. usambarae (E)	Usambara Shrew	
C. viaria (E)	Savanna Path Shrew	[41]
C. voi (E)	Voi Shrew	
C. whitakeri (P)	Whitaker's Shrew	
C. wimmeri (E)	Wimmer's Shrew	
C. xantippe (E)	Vermiculate Shrew	
C. yankariensis (E)	Yankari Shrew	[41]
C. zaphiri (E)	Zaphir's Shrew	
C. zarudnyi (P)	Zarudny's Shrew	
C. zimmeri (E)	Zimmer's Shrew	
C. zimmermanni (P)	Zimmermann's Shrew	
Genus *Diplomesodon*		
D. pulchellum (P)	Piebald Shrew	[16]
Genus *Feroculus*		
F. feroculus (Or)	Kelaart's Long-clawed Shrew	[16]
Genus *Myosorex*	Mouse Shrews	[16]
M. babaulti (E)	Babault's Mouse Shrew	
M. blarina (E)	Montane Mouse Shrew	
M. cafer (E)	Dark-footed Forest Shrew	[16]
M. eisentrauti (E)	Eisentraut's Mouse Shrew	
M. geata (E)	Geata Mouse Shrew	
M. longicaudatus (E)	Long-tailed Forest Shrew	[16]
M. okuensis (E)	Oku Mouse Shrew	
M. rumpii (E)	Rumpi Mouse Shrew	
M. schalleri (E)	Schaller's Mouse Shrew	
M. sclateri (E)	Sclater's Tiny Mouse Shrew	
M. tenuis (E)	Thin Mouse Shrew	
M. varius (E)	Forest Shrew	[16]

Genus *Paracrocidura*	African Shrews	
P. *graueri* (E)	Grauer's Shrew	
P. *maxima* (E)	Greater Shrew	
P. *schoutedeni* (E)	Schouteden's Shrew	
Genus *Ruwenzorisorex*		
R. *suncoides* (E)	Ruwenzori Shrew	
Genus *Scutisorex*		
S. *somereni* (E)	Armored Shrew	[16]
Genus *Solisorex*		
S. *pearsoni* (Or)	Pearson's Long-clawed Shrew	[16]
Genus *Suncus*	Pygmy and Dwarf Shrews	
S. *ater* (Or)	Black Shrew	[16]
S. *dayi* (Or)	Day's Shrew	[109]
S. *etruscus* (P, Or)	White-toothed Pygmy Shrew	[16]
S. *fellowesgordoni* (Or)	Sri Lanka Shrew	
S. *hosei* (Or)	Hose's Shrew	
S. *infinitesimus* (E)	Least Dwarf Shrew	[16]
S. *lixus* (E)	Greater Dwarf Shrew	[16]
S. *madagascariensis* (E)	Madagascan Shrew	
S. *malayanus* (Or)	Malayan Pygmy Shrew	
S. *mertensi* (A)	Flores Shrew	[20]
S. *montanus* (Or)	Sri Lanka Highland Shrew	
S. *murinus* (P, Or)	Asian House Shrew	[48]
S. *remyi* (E)	Remy's Shrew	
S. *stoliczkanus* (P, Or)	Anderson's Shrew	[17]
S. *varilla* (E)	Lesser Dwarf Shrew	[16]
S. *zeylanicus* (Or)	Jungle Shrew	
Genus *Surdisorex*	Kenyan Shrews	
S. *norae* (E)	Aberdare Shrew	
S. *polulus* (E)	Mt. Kenya Shrew	
Genus *Sylvisorex*	Forest Musk Shrews	[86]
S. *granti* (E)	Grant's Shrew	[109]
S. *howelli* (E)	Howell's Shrew	
S. *isabellae* (E)	Isabella Shrew	
S. *johnstoni* (E)	Johnston's Shrew	[109]
S. *lunaris* (E)	Crescent Shrew	
S. *megalura* (E)	Climbing Shrew	[16]
S. *morio* (E)	Arrogant Shrew	

S. ollula (E)	Forest Musk Shrew	
S. oriundus (E)	Mountain Shrew	
S. vulcanorum (E)	Volcano Shrew	
Subfamily Soricinae		
Genus *Anourosorex*		
A. squamipes (P, Or)	Mole Shrew	[16]
Genus *Blarina*	American Short-tailed Shrews	[16]
B. brevicauda (Nea)	Northern Short-tailed Shrew	[114]
B. carolinensis (Nea)	Southern Short-tailed Shrew	[114]
B. hylophaga (Nea)	Elliot's Short-tailed Shrew	[114]
Genus *Blarinella*	Asiatic Short-tailed Shrews	[109]
B. quadraticauda (P)	Sichuan Short-tailed Shrew	[17]
B. wardi (P, Or)	Ward's Short-tailed Shrew	
Genus *Chimarrogale*	Oriental Water Shrews	[16]
C. hantu (Or)	Hantu Water Shrew	
C. himalayica (P, Or)	Himalayan Water Shrew	[17]
C. phaeura (Or)	Sunda Water Shrew	[17]
C. platycephala (P)	Flat-headed Water Shrew	
C. styani (P, Or)	Styan's Water Shrew	[17]
C. sumatrana (Or)	Sumatra Water Shrew	
Genus *Cryptotis*	Small-eared Shrews	[16]
C. avia (Neo)	Andean Small-eared Shrew	
C. endersi (Neo)	Enders's Small-eared Shrew	[98]
C. goldmani (Nea, Neo)	Goldman's Small-eared Shrew	[98]
C. goodwini (Nea, Neo)	Goodwin's Small-eared Shrew	[98]
C. gracilis (Neo)	Talamancan Small-eared Shrew	[98]
C. hondurensis (Neo)	Honduran Small-eared Shrew	[98]
C. magna (Nea)	Big Small-eared Shrew	[16]
C. meridensis (Neo)	Merida Small-eared Shrew	
C. mexicana (Nea)	Mexican Small-eared Shrew	[98]
C. montivaga (Neo)	Ecuadorean Small-eared Shrew	
C. nigrescens (Nea, Neo)	Blackish Small-eared Shrew	[98]

C. parva (Nea, Neo)	Least Shrew	[114]
C. squamipes (Neo)	Scaly-footed Small-eared Shrew	
C. thomasi (Neo)	Thomas's Small-eared Shrew	
Genus *Megasorex*		
M. gigas (Nea)	Mexican Shrew	
Genus *Nectogale*		
N. elegans (P, Or)	Elegant Water Shrew	[16]
Genus *Neomys*	Old World Water Shrews	[86]
N. anomalus (P)	Southern Water Shrew	[16]
N. fodiens (P)	Eurasian Water Shrew	[16]
N. schelkovnikovi (P)	Transcaucasian Water Shrew	[16]
Genus *Notiosorex*		
N. crawfordi (Nea)	Desert Shrew	[114]
Genus *Sorex*	Holarctic Shrews	
S. alaskanus (Nea)	Glacier Bay Water Shrew	[54]
S. alpinus (P)	Alpine Shrew	[16]
S. araneus (P)	Eurasian Shrew	
S. arcticus (Nea)	Arctic Shrew	[114]
S. arizonae (Nea)	Arizona Shrew	[114]
S. asper (P)	Tien Shan Shrew	[16]
S. bairdii (Nea)	Baird's Shrew	[114]
S. bedfordiae (P, Or)	Lesser Striped Shrew	[16]
S. bendirii (Nea)	Marsh Shrew	[114]
S. buchariensis (P)	Pamir Shrew	[16]
S. caecutiens (P)	Laxmann's Shrew	[16]
S. camtschatica (P)	Kamchatka Shrew	
S. cansulus (P)	Gansu Shrew	
S. cinereus (Nea)	Cinereus Shrew	[114]
S. coronatus (P)	Crowned Shrew	
S. cylindricauda (P)	Stripe-backed Shrew	[109]
S. daphaenodon (P)	Large-toothed Siberian Shrew	[16]
S. dispar (Nea)	Long-tailed Shrew	[114]
S. emarginatus (Nea)	Zacatecas Shrew	[72]
S. excelsus (P)	Lofty Shrew	
S. fumeus (Nea)	Smoky Shrew	[114]

S. *gaspensis* (Nea)	Gaspé Shrew	[114]
S. *gracillimus* (P)	Slender Shrew	[16]
S. *granarius* (P)	Lagranja Shrew	
S. *haydeni* (Nea)	Prairie Shrew	[114]
S. *hosonoi* (P)	Azumi Shrew	[16]
S. *hoyi* (Nea)	Pygmy Shrew	[114]
S. *hydrodromus* (Nea)	Pribilof Island Shrew	[114]
S. *isodon* (P)	Even-toothed Shrew	
S. *jacksoni* (Nea)	St. Lawrence Island Shrew	[109]
S. *kozlovi* (P)	Kozlov's Shrew	
S. *leucogaster* (P)	Paramushir Shrew	
S. *longirostris* (Nea)	Southeastern Shrew	[114]
S. *lyelli* (Nea)	Mt. Lyell Shrew	[114]
S. *macrodon* (Nea)	Large-toothed Shrew	[16]
S. *merriami* (Nea)	Merriam's Shrew	[114]
S. *milleri* (Nea)	Carmen Mountain Shrew	[16]
S. *minutissimus* (P)	Miniscule Shrew	
S. *minutus* (P)	Eurasian Pygmy Shrew	[16]
S. *mirabilis* (P)	Ussuri Shrew	
S. *monticolus* (Nea)	Montane Shrew	[114]
S. *nanus* (Nea)	Dwarf Shrew	[114]
S. *oreopolus* (Nea)	Mexican Long-tailed Shrew	[16]
S. *ornatus* (Nea)	Ornate Shrew	[114]
S. *pacificus* (Nea)	Pacific Shrew	[114]
S. *palustris* (Nea)	Water Shrew	[114]
S. *planiceps* (P, Or)	Kashmir Shrew	[16]
S. *portenkoi* (P)	Portenko's Shrew	
S. *preblei* (Nea)	Preble's Shrew	[114]
S. *raddei* (P)	Radde's Shrew	[16]
S. *roboratus* (P)	Flat-skulled Shrew	[16]
S. *sadonis* (P)	Sado Shrew	
S. *samniticus* (P)	Apennine Shrew	[16]
S. *satunini* (P)	Caucasian Shrew	[109]
S. *saussurei* (Nea, Neo)	Saussure's Shrew	[16]
S. *sclateri* (Nea)	Sclater's Shrew	[98]
S. *shinto* (P)	Shinto Shrew	
S. *sinalis* (P)	Chinese Shrew	
S. *sonomae* (Nea)	Fog Shrew	[114]

S. stizodon (Nea)	San Cristobal Shrew	[98]
S. tenellus (Nea)	Inyo Shrew	[114]
S. thibetanus (P, Or)	Tibetan Shrew	
S. trowbridgii (Nea)	Trowbridge's Shrew	[114]
S. tundrensis (Nea, P)	Tundra Shrew	[114]
S. ugyunak (Nea)	Barren Ground Shrew	[114]
S. unguiculatus (P)	Long-clawed Shrew	[16]
S. vagrans (Nea)	Vagrant Shrew	[114]
S. ventralis (Nea)	Chestnut-bellied Shrew	
S. veraepacis (Nea, Neo)	Verapaz Shrew	[98]
S. volnuchini (P)	Caucasian Pygmy Shrew	[16]
Genus *Soriculus*	Asiatic Shrews	[86]
S. caudatus (P, Or)	Hodgson's Brown-toothed Shrew	[16]
S. fumidus (P)	Taiwan Brown-toothed Shrew	
S. hypsibius (P)	De Winton's Shrew	[16]
S. lamula (P)	Lamulate Shrew	
S. leucops (P, Or)	Long-tailed Brown-toothed Shrew	
S. macrurus (P, Or)	Long-tailed Mountain Shrew	
S. nigrescens (P, Or)	Himalayan Shrew	[16]
S. parca (P, Or)	Lowe's Shrew	[109]
S. salenskii (P)	Salenski's Shrew	[16]
S. smithii (P)	Smith's Shrew	[16]
Family Talpidae	Desmans, Moles, and Shrew Moles	[16]
Subfamily Desmaninae		
Genus *Desmana*		
D. moschata (P)	Russian Desman	[16]
Genus *Galemys*		
G. pyrenaicus (P)	Pyrenean Desman	[16]
Subfamily Talpinae		
Genus *Condylura*		
C. cristata (Nea)	Star-nosed Mole	[114]
Genus *Euroscaptor*	Oriental Moles	
E. grandis (P, Or)	Greater Chinese Mole	
E. klossi (Or)	Kloss's Mole	
E. longirostris (P)	Long-nosed Mole	
E. micrura (P, Or)	Himalayan Mole	[16]

E. mizura (P)	Japanese Mountain Mole	[16]
E. parvidens (Or)	Small-toothed Mole	
Genus *Mogera*	East Asian Moles	[86]
M. etigo (P)	Echigo Mole	
M. insularis (P)	Insular Mole	
M. kobeae (P)	Kobe Mole	
M. minor (P)	Small Japanese Mole	
M. robusta (P)	Large Mole	
M. tokudae (P)	Tokuda's Mole	
M. wogura (P)	Japanese Mole	[16]
Genus *Nesoscaptor*		
N. uchidai (P)	Ryukyu Mole	
Genus *Neurotrichus*		
N. gibbsii (Nea)	American Shrew Mole	[114]
Genus *Parascalops*		
P. breweri (Nea)	Hairy-tailed Mole	[114]
Genus *Parascaptor*		
P. leucura (P, Or)	White-tailed Mole	
Genus *Scalopus*		
S. aquaticus (Nea)	Eastern Mole	[11]
Genus *Scapanulus*		
S. oweni (P)	Gansu Mole	
Genus *Scapanus*	Western American Moles	[86]
S. latimanus (Nea)	Broad-footed Mole	[114]
S. orarius (Nea)	Coast Mole	[114]
S. townsendii (Nea)	Townsend's Mole	[114]
Genus *Scaptochirus*		
S. moschatus (P)	Short-faced Mole	[16]
Genus *Scaptonyx*		
S. fusicaudus (P, Or)	Long-tailed Mole	[16]
Genus *Talpa*	Old World Moles	[86]
T. altaica (P)	Siberian Mole	[16]
T. caeca (P)	Mediterranean Mole	[16]
T. caucasica (P)	Caucasian Mole	[16]
T. europaea (P)	European Mole	[16]
T. levantis (P)	Levantine Mole	
T. occidentalis (P)	Iberian Mole	
T. romana (P)	Roman Mole	[16]
T. stankovici (P)	Stankovic's Mole	
T. streeti (P)	Persian Mole	[16]

Genus *Urotrichus*	Japanese Shrew Moles	[86]
U. pilirostris (P)	True's Shrew Mole	[109]
U. talpoides (P)	Japanese Shrew Mole	[109]
Subfamily Uropsilinae		
Genus *Uropsilus*	Asiatic Shrew Moles	[86]
U. andersoni (P)	Anderson's Shrew Mole	
U. gracilis (P, Or)	Gracile Shrew Mole	
U. investigator (P)	Inquisitive Shrew Mole	
U. soricipes (P)	Chinese Shrew Mole	[16]

ORDER SCANDENTIA	Tree Shrews	[86]
Family Tupaiidae		
Subfamily Tupaiinae		
Genus *Anathana*		
A. ellioti (Or)	Madras Tree Shrew	[17]
Genus *Dendrogale*	Smooth-tailed Tree Shrews	
D. melanura (Or)	Bornean Smooth-tailed Tree Shrew	[16]
D. murina (Or)	Northern Smooth-tailed Tree Shrew	[16]
Genus *Tupaia*	Tree Shrews	[86]
T. belangeri (P, Or)	Northern Tree Shrew	[16]
T. chrysogaster (Or)	Golden-bellied Tree Shrew	
T. dorsalis (Or)	Striped Tree Shrew	[16]
T. glis (Or)	Common Tree Shrew	[16]
T. gracilis (Or)	Slender Tree Shrew	[16]
T. javanica (Or)	Javan Tree Shrew	[16]
T. longipes (Or)	Long-footed Tree Shrew	
T. minor (Or)	Pygmy Tree Shrew	[16]
T. montana (Or)	Mountain Tree Shrew	[16]
T. nicobarica (Or)	Nicobar Tree Shrew	[16]
T. palawanensis (Or)	Palawan Tree Shrew	[48]
T. picta (Or)	Painted Tree Shrew	[16]
T. splendidula (Or)	Ruddy Tree Shrew	[16]
T. tana (Or)	Large Tree Shrew	[16]
Genus *Urogale*		
U. everetti (Or)	Mindanao Tree Shrew	[48]
Subfamily Ptilocercinae		
Genus *Ptilocercus*		
P. lowii (Or)	Pen-tailed Tree Shrew	[16]

ORDER DERMOPTERA   Flying Lemurs   [16]
Family Cynocephalidae
  Genus *Cynocephalus*
    C. variegatus (Or)   Malayan Flying Lemur   [16]
    C. volans (Or)   Philippine Flying Lemur   [16]

ORDER CHIROPTERA   Bats   [16]
Family Pteropodidae   Old World Fruit Bats   [86]
  Subfamily Pteropodinae
    Genus *Acerodon*   Island Fruit Bats   [109]
      A. celebensis (A)   Sulawesi Fruit Bat   [75]
      A. humilis (A)   Talaud Fruit Bat   [75]
      A. jubatus (Or)   Golden-capped Fruit Bat   [75]
      A. leucotis (Or)   Palawan Fruit Bat
      A. lucifer (Or)   Panay Golden-capped Fruit Bat
      A. mackloti (A)   Sunda Fruit Bat   [75]
    Genus *Aethalops*
      A. alecto (Or)   Pygmy Fruit Bat   [75]
    Genus *Alionycteris*
      A. paucidentata (Or)   Mindanao Pygmy Fruit Bat   [48]
    Genus *Aproteles*
      A. bulmerae (A)   Bulmer's Fruit Bat   [28]
    Genus *Balionycteris*
      B. maculata (Or)   Spotted-winged Fruit Bat   [75]
    Genus *Boneia*
      B. bidens (A)   Manado Fruit Bat
    Genus *Casinycteris*
      C. argynnis (E)   Short-palated Fruit Bat   [75]
    Genus *Chironax*
      C. melanocephalus (Or, A)   Black-capped Fruit Bat   [75]
    Genus *Cynopterus*   Short-nosed Fruit Bats   [86]
      C. brachyotis (Or, A)   Lesser Short-nosed Fruit Bat   [115]
      C. horsfieldi (Or, A)   Horsfield's Fruit Bat   [75]
      C. nusatenggara (A)   Nusatenggara Short-nosed Fruit Bat
      C. sphinx (P, Or)   Greater Short-nosed Fruit Bat   [115]
      C. titthaecheileus (Or, A)   Indonesian Short-nosed Fruit Bat

Genus *Dobsonia*	Naked-backed Fruit Bats	[16]
D. *beauforti* (A)	Beaufort's Naked-backed Fruit Bat	
D. *chapmani* (Or)	Negros Naked-backed Fruit Bat	[48]
D. *emersa* (A)	Biak Naked-backed Fruit Bat	
D. *exoleta* (A)	Sulawesi Naked-backed Fruit Bat	[75]
D. *inermis* (A)	Solomons Naked-backed Fruit Bat	[75]
D. *minor* (A)	Lesser Naked-backed Fruit Bat	[75]
D. *moluccensis* (A)	Moluccan Naked-backed Fruit Bat	[27]
D. *pannietensis* (A)	Panniet Naked-backed Fruit Bat	
D. *peroni* (A)	Western Naked-backed Fruit Bat	[75]
D. *praedatrix* (A)	New Britain Naked-backed Fruit Bat	[75]
D. *viridis* (A)	Greenish Naked-backed Fruit Bat	[75]
Genus *Dyacopterus*		
D. *spadiceus* (Or)	Dyak Fruit Bat	[48]
Genus *Eidolon*	Eidolon Fruit Bats	[109]
E. *dupreanum* (E)	Madagascan Fruit Bat	
E. *helvum* (P, E)	Straw-colored Fruit Bat	[75]
Genus *Epomophorus*	Epauletted Fruit Bats	[16]
E. *angolensis* (E)	Angolan Epauletted Fruit Bat	[75]
E. *gambianus* (E)	Gambian Epauletted Fruit Bat	[16]
E. *grandis* (E)	Lesser Angolan Epauletted Fruit Bat	[75]
E. *labiatus* (P, E)	Ethiopian Epauletted Fruit Bat	[75]
E. *minimus* (E)	East African Epauletted Fruit Bat	[75]
E. *wahlbergi* (E)	Wahlberg's Epauletted Fruit Bat	[75]

Genus *Epomops* — African Epauletted Bats
    E. buettikoferi (E) — Buettikofer's Epauletted Bat [75]
    E. dobsoni (E) — Dobson's Fruit Bat [44]
    E. franqueti (E) — Franquet's Epauletted Bat [75]
Genus *Haplonycteris*
    H. fischeri (Or) — Philippine Pygmy Fruit Bat [48]
Genus *Harpyionycteris* — Harpy Fruit Bats [86]
    H. celebensis (A) — Sulawesi Harpy Fruit Bat
    H. whiteheadi (Or) — Harpy Fruit Bat [48]
Genus *Hypsignathus*
    H. monstrosus (E) — Hammer-headed Fruit Bat [75]
Genus *Latidens*
    L. salimalii (Or) — Salim Ali's Fruit Bat [8]
Genus *Megaerops* — Tailless Fruit Bats [109]
    M. ecaudatus (Or) — Temminck's Tailless Fruit Bat
    M. kusnotoi (Or) — Javan Tailless Fruit Bat [75]
    M. niphanae (Or) — Ratanaworabhan's Fruit Bat [8]
    M. wetmorei (Or) — White-collared Fruit Bat [91]
Genus *Micropteropus* — Dwarf Epauletted Fruit Bats [16]
    M. intermedius (E) — Hayman's Dwarf Epauletted Fruit Bat
    M. pusillus (E) — Peters's Dwarf Epauletted Fruit Bat [75]
Genus *Myonycteris* — Little Collared Fruit Bats [86]
    M. brachycephala (E) — Sao Tomé Collared Fruit Bat [44]
    M. relicta (E) — East African Little Collared Fruit Bat
    M. torquata (E) — Little Collared Fruit Bat [75]
Genus *Nanonycteris*
    N. veldkampi (E) — Veldkamp's Bat [103]
Genus *Neopteryx*
    N. frosti (A) — Small-toothed Fruit Bat [75]
Genus *Nyctimene* — Tube-nosed Fruit Bats [16]
    N. aello (A) — Broad-striped Tube-nosed Fruit Bat [75]
    N. albiventer (A) — Common Tube-nosed Fruit Bat [75]
    N. celaeno (A) — Dark Tube-nosed Fruit Bat

N. cephalotes (A)	Pallas's Tube-nosed Fruit Bat	[75]
N. certans (A)	Mountain Tube-nosed Fruit Bat	[28]
N. cyclotis (A)	Round-eared Tube-nosed Fruit Bat	[28]
N. draconilla (A)	Dragon Tube-nosed Fruit Bat	
N. major (A)	Island Tube-nosed Fruit Bat	[29]
N. malaitensis (A)	Malaita Tube-nosed Fruit Bat	[75]
N. masalai (A)	Demonic Tube-nosed Fruit Bat	
N. minutus (A)	Lesser Tube-nosed Fruit Bat	[16]
N. rabori (Or)	Philippine Tube-nosed Fruit Bat	[48]
N. robinsoni (A)	Queensland Tube-nosed Fruit Bat	[75]
N. sanctacrucis (A)	Nendo Tube-nosed Fruit Bat	[29]
N. vizcaccia (A)	Umboi Tube-nosed Fruit Bat	
Genus *Otopteropus*		
O. cartilagonodus (Or)	Luzon Fruit Bat	[20]
Genus *Paranyctimene*		
P. raptor (A)	Unstriped Tube-nosed Bat	[28]
Genus *Penthetor*		
P. lucasi (Or)	Lucas's Short-nosed Fruit Bat	[75]
Genus *Plerotes*		
P. anchietai (E)	d'Anchieta's Fruit Bat	[75]
Genus *Ptenochirus*	Musky Fruit Bats	
P. jagori (Or)	Greater Musky Fruit Bat	
P. minor (Or)	Lesser Musky Fruit Bat	[48]
Genus *Pteralopex*	Monkey-faced Bats	
P. acrodonta (Oc)	Fijian Monkey-faced Bat	[29]
P. anceps (A)	Bougainville Monkey-faced Bat	[29]
P. atrata (A)	Guadalcanal Monkey-faced Bat	[29]
P. pulchra (A)	Montane Monkey-faced Bat	[29]

Genus *Pteropus*	Flying Foxes	[16]
P. *admiralitatum* (A)	Admiralty Flying Fox	[29]
P. *aldabrensis* (Oc)	Aldabra Flying Fox	[44]
P. *alecto* (Or, A)	Black Flying Fox	[110]
P. *anetianus* (Oc)	Vanauatu Flying Fox	[29]
P. *argentatus* (A)	Ambon Flying Fox	[29]
P. *brunneus* (A)	Dusky Flying Fox	[110]
P. *caniceps* (A)	North Moluccan Flying Fox	[29]
P. *chrysoproctus* (A)	Moluccan Flying Fox	[29]
P. *conspicillatus* (A)	Spectacled Flying Fox	[75]
P. *dasymallus* (P)	Ryukyu Flying Fox	[48]
P. *faunulus* (Or)	Nicobar Flying Fox	[8]
P. *fundatus* (Oc)	Banks Flying Fox	[29]
P. *giganteus* (P, Or)	Indian Flying Fox	[75]
P. *gilliardi* (A)	Gilliard's Flying Fox	[75]
P. *griseus* (Or, A)	Gray Flying Fox	[75]
P. *howensis* (A)	Ontong Java Flying Fox	[29]
P. *hypomelanus* (Oc, Or, A)	Variable Flying Fox	[29]
P. *insularis* (Oc)	Ruck Flying Fox	[29]
P. *leucopterus* (Or)	White-winged Flying Fox	[48]
P. *livingstonei* (E)	Comoro Black Flying Fox	[16]
P. *lombocensis* (A)	Lombok Flying Fox	[16]
P. *lylei* (Or)	Lyle's Flying Fox	[16]
P. *macrotis* (A)	Big-eared Flying Fox	[28]
P. *mahaganus* (A)	Sanborn's Flying Fox	[29]
P. *mariannus* (Oc)	Marianas Flying Fox	[29]
P. *mearnsi* (Or)	Mearns's Flying Fox	[20]
P. *melanopogon* (A)	Black-bearded Flying Fox	[29]
P. *melanotus* (Or)	Black-eared Flying Fox	[20]
P. *molossinus* (Oc)	Caroline Flying Fox	[29]
P. *neohibernicus* (A)	Great Flying Fox	[29]
P. *niger* (Oc)	Greater Mascarene Flying Fox	[44]
P. *nitendiensis* (A)	Temotu Flying Fox	[29]
P. *ocularis* (A)	Ceram Flying Fox	[29]
P. *ornatus* (Oc)	Ornate Flying Fox	[29]
P. *personatus* (A)	Masked Flying Fox	[29]
P. *phaeocephalus* (Oc)	Mortlock Flying Fox	[29]
P. *pilosus* (Oc)	Large Palau Flying Fox	[29]
P. *pohlei* (A)	Geelvink Bay Flying Fox	[29]

P. poliocephalus (A)	Gray-headed Flying Fox	[110]
P. pselaphon (P)	Bonin Flying Fox	
P. pumilus (Or)	Little Golden-mantled Flying Fox	[48]
P. rayneri (A)	Solomons Flying Fox	[29]
P. rodricensis (Oc)	Rodriguez Flying Fox	[44]
P. rufus (E)	Madagascan Flying Fox	[16]
P. samoensis (Oc)	Samoan Flying Fox	[29]
P. sanctacrucis (A)	Santa Cruz Flying Fox	
P. scapulatus (A)	Little Red Flying Fox	[28]
P. seychellensis (Oc, E)	Seychelles Flying Fox	[16]
P. speciosus (Or)	Philippine Gray Flying Fox	[48]
P. subniger (Oc)	Dark Flying Fox	[86]
P. temmincki (A)	Temminck's Flying Fox	[29]
P. tokudae (Oc)	Guam Flying Fox	[29]
P. tonganus (Oc, A)	Pacific Flying Fox	[29]
P. tuberculatus (A)	Vanikoro Flying Fox	[29]
P. vampyrus (Or, A)	Large Flying Fox	[8]
P. vetulus (Oc)	New Caledonia Flying Fox	[29]
P. voeltzkowi (E)	Pemba Flying Fox	[16]
P. woodfordi (A)	Dwarf Flying Fox	[29]
Genus *Rousettus*	Rousette Fruit Bats	[86]
R. aegyptiacus (P, E)	Egyptian Rousette	[16]
R. amplexicaudatus (Or, A)	Geoffroy's Rousette	[75]
R. angolensis (E)	Angolan Rousette	
R. celebensis (A)	Sulawesi Rousette	[75]
R. lanosus (E)	Long-haired Rousette	
R. leschenaulti (P, Or)	Leschenault's Rousette	[16]
R. madagascariensis (E)	Madagascan Rousette	[75]
R. obliviosus (E)	Comoro Rousette	[75]
R. spinalatus (Or)	Bare-backed Rousette	[91]
Genus *Scotonycteris*	West African Fruit Bats	
S. ophiodon (E)	Pohle's Fruit Bat	[75]
S. zenkeri (E)	Zenker's Fruit Bat	[75]
Genus *Sphaerias*		
S. blanfordi (P, Or)	Blanford's Fruit Bat	[75]
Genus *Styloctenium*		
S. wallacei (A)	Stripe-faced Fruit Bat	[75]
Genus *Thoopterus*		
T. nigrescens (A)	Swift Fruit Bat	[75]

Subfamily Macroglossinae

Genus *Eonycteris*	Dawn Bats	[86]
E. major (Or)	Greater Dawn Bat	
E. spelaea (P, Or, A)	Lesser Dawn Bat	
Genus *Macroglossus*	Long-tongued Fruit Bats	[16]
M. minimus (Or, A)	Lesser Long-tongued Fruit Bat	[67]
M. sobrinus (Or)	Greater Long-tongued Fruit Bat	[67]
Genus *Megaloglossus*		
M. woermanni (E)	Woermann's Bat	[75]
Genus *Melonycteris*	Black-bellied Fruit Bats	[109]
M. aurantius (A)	Orange Fruit Bat	[75]
M. melanops (A)	Black-bellied Fruit Bat	[75]
M. woodfordi (A)	Woodford's Fruit Bat	[75]
Genus *Notopteris*		
N. macdonaldi (Oc)	Long-tailed Fruit Bat	[75]
Genus *Syconycteris*	Blossom Bats	[86]
S. australis (A)	Southern Blossom Bat	[16]
S. carolinae (A)	Halmahera Blossom Bat	[29]
S. hobbit (A)	Moss-forest Blossom Bat	[28]
Family Rhinopomatidae	Mouse-tailed Bats	[16]
Genus *Rhinopoma*		
R. hardwickei (P, E, Or)	Lesser Mouse-tailed Bat	[8]
R. microphyllum (P, E, Or)	Greater Mouse-tailed Bat	[8]
R. muscatellum (P)	Small Mouse-tailed Bat	[94]
Family Craseonycteridae		
Genus *Craseonycteris*		
C. thonglongyai (Or)	Hog-nosed Bat	[16]
Family Emballonuridae	Sheath-tailed Bats	[86]
Genus *Balantiopteryx*	Least Sac-winged Bats	[16]
B. infusca (Neo)	Ecuadorian Sac-winged Bat	[2]
B. io (Nea, Neo)	Thomas's Sac-winged Bat	[2]
B. plicata (Nea, Neo)	Gray Sac-winged Bat	[98]
Genus *Centronycteris*		
C. maximiliani (Nea, Neo)	Shaggy Bat	[26]
Genus *Coleura*	Peters's Sheath-tailed Bats	[103]
C. afra (P, E)	African Sheath-tailed Bat	[16]
C. seychellensis (Oc)	Seychelles Sheath-tailed Bat	[16]

Genus *Cormura*
    C. brevirostris (Neo)   Chestnut Sac-winged Bat   [98]
Genus *Cyttarops*
    C. alecto (Neo)   Short-eared Bat   [38]
Genus *Diclidurus*   Ghost Bats   [86]
    D. albus (Nea, Neo)   Northern Ghost Bat   [98]
    D. ingens (Neo)   Greater Ghost Bat
    D. isabellus (Neo)   Isabelle's Ghost Bat
    D. scutatus (Neo)   Lesser Ghost Bat
Genus *Emballonura*   Old World Sheath-tailed Bats   [16]
    E. alecto (Or, A)   Small Asian Sheath-tailed Bat   [29]
    E. atrata (E)   Peters's Sheath-tailed Bat   [16]
    E. beccarii (A)   Beccari's Sheath-tailed Bat   [16]
    E. dianae (A)   Large-eared Sheath-tailed Bat   [28]
    E. furax (A)   New Guinea Sheath-tailed Bat   [28]
    E. monticola (Or, A)   Lesser Sheath-tailed Bat   [16]
    E. raffrayana (A)   Raffray's Sheath-tailed Bat   [16]
    E. semicaudata (Oc)   Polynesian Sheath-tailed Bat   [29]
Genus *Mosia*
    M. nigrescens (A)   Dark Sheath-tailed Bat
Genus *Peropteryx*   Dog-like Bats   [86]
    P. kappleri (Nea, Neo)   Greater Dog-like Bat   [98]
    P. leucoptera (Neo)   White-winged Dog-like Bat
    P. macrotis (Nea, Neo)   Lesser Dog-like Bat   [39]
Genus *Rhynchonycteris*
    R. naso (Nea, Neo)   Proboscis Bat   [98]
Genus *Saccolaimus*   Pouched Bats
    S. flaviventris (A)   Yellow-bellied Pouched Bat   [16]
    S. mixtus (A)   Troughton's Pouched Bat   [16]
    S. peli (E)   Pel's Pouched Bat   [16]
    S. pluto (Or)   Philippine Pouched Bat
    S. saccolaimus (Or, A)   Naked-rumped Pouched Bat
Genus *Saccopteryx*   Sac-winged Bats
    S. bilineata (Nea, Neo)   Greater Sac-winged Bat   [34]
    S. canescens (Neo)   Frosted Sac-winged Bat

S. gymnura (Neo)	Amazonian Sac-winged Bat	
S. leptura (Nea, Neo)	Lesser Sac-winged Bat	[34]
Genus *Taphozous*	Tomb Bats	[86]
T. australis (A)	Coastal Tomb Bat	
T. georgianus (A)	Sharp-nosed Tomb Bat	
T. hamiltoni (E)	Hamilton's Tomb Bat	[16]
T. hildegardeae (E)	Hildegarde's Tomb Bat	[16]
T. hilli (A)	Hill's Tomb Bat	
T. kapalgensis (A)	Arnhem Tomb Bat	
T. longimanus (Or, A)	Long-winged Tomb Bat	[16]
T. mauritianus (Oc, E)	Mauritian Tomb Bat	[16]
T. melanopogon (P, Or, A)	Black-bearded Tomb Bat	[48]
T. nudiventris (P, E, Or)	Naked-rumped Tomb Bat	[16]
T. perforatus (P, E, Or)	Egyptian Tomb Bat	[16]
T. philippinensis (Or)	Philippine Tomb Bat	
T. theobaldi (Or, A)	Theobald's Tomb Bat	[16]
Family Nycteridae	Slit-faced Bats	[16]
Genus *Nycteris*		
N. arge (E)	Bate's Slit-faced Bat	[16]
N. gambiensis (E)	Gambian Slit-faced Bat	[16]
N. grandis (E)	Large Slit-faced Bat	[16]
N. hispida (E)	Hairy Slit-faced Bat	[16]
N. intermedia (E)	Intermediate Slit-faced Bat	
N. javanica (Or)	Javan Slit-faced Bat	[16]
N. macrotis (E)	Large-eared Slit-faced Bat	[44]
N. major (E)	Ja Slit-faced Bat	[16]
N. nana (E)	Dwarf Slit-faced Bat	[16]
N. thebaica (P, E)	Egyptian Slit-faced Bat	[16]
N. tragata (Or)	Malayan Slit-faced Bat	
N. woodi (E)	Wood's Slit-faced Bat	[16]
Family Megadermatidae	False Vampire Bats	[16]
Genus *Cardioderma*		
C. cor (E)	Heart-nosed Bat	[16]
Genus *Lavia*		
L. frons (E)	Yellow-winged Bat	[16]
Genus *Macroderma*		
M. gigas (A)	Australian False Vampire Bat	[16]
Genus *Megaderma*	Asian False Vampire Bats	[86]
M. lyra (P, Or)	Greater False Vampire Bat	[16]
M. spasma (Or, A)	Lesser False Vampire Bat	[48]

Family Rhinolophidae	Horseshoe Bats	[16]
Subfamily Rhinolophinae		
Genus *Rhinolophus*	Horseshoe Bats	[86]
R. *acuminatus* (Or)	Acuminate Horseshoe Bat	[48]
R. *adami* (E)	Adam's Horseshoe Bat	
R. *affinis* (P, Or, A)	Intermediate Horseshoe Bat	[16]
R. *alcyone* (E)	Halcyon Horseshoe Bat	[16]
R. *anderseni* (Or)	Andersen's Horseshoe Bat	[48]
R. *arcuatus* (Or, A)	Arcuate Horseshoe Bat	[48]
R. *blasii* (P, E)	Blasius's Horseshoe Bat	[16]
R. *borneensis* (Or)	Bornean Horseshoe Bat	[16]
R. *canuti* (Or, A)	Canut's Horseshoe Bat	
R. *capensis* (E)	Cape Horseshoe Bat	[16]
R. *celebensis* (Or, A)	Sulawesi Horseshoe Bat	[16]
R. *clivosus* (P, E)	Geoffroy's Horseshoe Bat	[16]
R. *coelophyllus* (Or)	Croslet Horseshoe Bat	[16]
R. *cognatus* (Or)	Andaman Horseshoe Bat	[8]
R. *cornutus* (P)	Little Japanese Horseshoe Bat	[16]
R. *creaghi* (Or, A)	Creagh's Horseshoe Bat	[16]
R. *darlingi* (E)	Darling's Horseshoe Bat	[16]
R. *deckenii* (E)	Decken's Horseshoe Bat	[16]
R. *denti* (E)	Dent's Horseshoe Bat	[16]
R. *eloquens* (E)	Eloquent Horseshoe Bat	
R. *euryale* (P)	Mediterranean Horseshoe Bat	[16]
R. *euryotis* (A)	Broad-eared Horseshoe Bat	[16]
R. *ferrumequinum* (P, Or)	Greater Horseshoe Bat	[16]
R. *fumigatus* (E)	Rüppell's Horseshoe Bat	[16]
R. *guineensis* (E)	Guinean Horseshoe Bat	
R. *hildebrandti* (E)	Hildebrandt's Horseshoe Bat	[16]
R. *hipposideros* (P, E, Or)	Lesser Horseshoe Bat	[16]
R. *imaizumii* (P)	Imaizumi's Horseshoe Bat	
R. *inops* (Or)	Philippine Forest Horseshoe Bat	[48]
R. *keyensis* (A)	Insular Horseshoe Bat	[16]
R. *landeri* (E)	Lander's Horseshoe Bat	[16]
R. *lepidus* (P, Or)	Blyth's Horseshoe Bat	[16]
R. *luctus* (P, Or)	Woolly Horseshoe Bat	[16]

R. *maclaudi* (E)	Maclaud's Horseshoe Bat	[16]
R. *macrotis* (P, Or)	Big-eared Horseshoe Bat	[48]
R. *malayanus* (Or)	Malayan Horseshoe Bat	[16]
R. *marshalli* (Or)	Marshall's Horseshoe Bat	[16]
R. *megaphyllus* (Or, A)	Smaller Horseshoe Bat	[17]
R. *mehelyi* (P)	Mehely's Horseshoe Bat	[16]
R. *mitratus* (Or)	Mitred Horseshoe Bat	[8]
R. *monoceros* (P)	Formosan Lesser Horse-	
	shoe Bat	[68]
R. *nereis* (Or)	Neriad Horseshoe Bat	
R. *osgoodi* (P)	Osgood's Horseshoe Bat	
R. *paradoxolophus* (Or)	Bourret's Horseshoe Bat	[16]
R. *pearsoni* (P, Or)	Pearson's Horseshoe Bat	[16]
R. *philippinensis* (Or, A)	Large-eared Horseshoe Bat	[110]
R. *pusillus* (P, Or)	Least Horseshoe Bat	[16]
R. *rex* (P)	King Horseshoe Bat	
R. *robinsoni* (Or)	Peninsular Horseshoe Bat	[16]
R. *rouxi* (P, Or)	Rufous Horseshoe Bat	[27]
R. *rufus* (Or)	Large Rufous Horseshoe	
	Bat	[48]
R. *sedulus* (Or)	Lesser Woolly Horseshoe	
	Bat	[16]
R. *shameli* (Or)	Shamel's Horseshoe Bat	[16]
R. *silvestris* (E)	Forest Horseshoe Bat	
R. *simplex* (A)	Lombok Horseshoe Bat	[16]
R. *simulator* (E)	Bushveld Horseshoe Bat	[16]
R. *stheno* (Or)	Lesser Brown Horseshoe	
	Bat	[16]
R. *subbadius* (Or)	Little Nepalese Horseshoe	
	Bat	[8]
R. *subrufus* (Or)	Small Rufous Horseshoe	
	Bat	[48]
R. *swinnyi* (E)	Swinny's Horseshoe Bat	[16]
R. *thomasi* (P, Or)	Thomas's Horseshoe Bat	[16]
R. *trifoliatus* (Or)	Trefoil Horseshoe Bat	[16]
R. *virgo* (Or)	Yellow-faced Horseshoe Bat	[48]
R. *yunanensis* (P, Or)	Dobson's Horseshoe Bat	[16]

Subfamily Hipposiderinae
    Genus *Anthops*

A. *ornatus* (A)	Flower-faced Bat	[16]

Genus *Asellia*	Trident Leaf-nosed Bats	[86]
A. patrizii (P, E)	Patrizi's Trident Leaf-nosed Bat	
A. tridens (P, E)	Trident Leaf-nosed Bat	[43]
Genus *Aselliscus*	Tate's Trident-nosed Bats	[86]
A. stoliczkanus (P, Or)	Stoliczka's Trident Bat	[16]
A. tricuspidatus (A)	Temminck's Trident Bat	[17]
Genus *Cloeotis*		
C. percivali (E)	Percival's Trident Bat	[16]
Genus *Coelops*	Tailless Leaf-nosed Bats	[109]
C. frithi (P, Or)	East Asian Tailless Leaf-nosed Bat	
C. hirsutus (Or)	Philippine Tailless Leaf-nosed Bat	
C. robinsoni (Or)	Malayan Tailless Leaf-nosed Bat	[16]
Genus *Hipposideros*	Roundleaf Bats	[91]
H. abae (E)	Aba Roundleaf Bat	
H. armiger (P, Or)	Great Roundleaf Bat	[115]
H. ater (Or, A)	Dusky Roundleaf Bat	[91]
H. beatus (E)	Benito Roundleaf Bat	
H. bicolor (Or, A)	Bicolored Roundleaf Bat	[91]
H. breviceps (Or)	Short-headed Roundleaf Bat	
H. caffer (P, E)	Sundevall's Roundleaf Bat	
H. calcaratus (A)	Spurred Roundleaf Bat	
H. camerunensis (E)	Greater Roundleaf Bat	
H. cervinus (Or, A)	Fawn Roundleaf Bat	[91]
H. cineraceus (P, Or)	Ashy Roundleaf Bat	[91]
H. commersoni (E)	Commerson's Roundleaf Bat	
H. coronatus (Or)	Large Mindanao Roundleaf Bat	[48]
H. corynophyllus (A)	Telefomin Roundleaf Bat	
H. coxi (Or)	Cox's Roundleaf Bat	[91]
H. crumeniferus (A)	Timor Roundleaf Bat	
H. curtus (E)	Short-tailed Roundleaf Bat	
H. cyclops (E)	Cyclops Roundleaf Bat	
H. diadema (Or, A)	Diadem Roundleaf Bat	[91]
H. dinops (A)	Fierce Roundleaf Bat	

H. doriae (Or)	Borneo Roundleaf Bat	
H. dyacorum (Or)	Dayak Roundleaf Bat	[91]
H. fuliginosus (E)	Sooty Roundleaf Bat	
H. fulvus (P, Or)	Fulvus Roundleaf Bat	[115]
H. galeritus (Or, A)	Cantor's Roundleaf Bat	[91]
H. halophyllus (Or)	Thailand Roundleaf Bat	
H. inexpectatus (A)	Crested Roundleaf Bat	
H. jonesi (E)	Jones's Roundleaf Bat	
H. lamottei (E)	Lamotte's Roundleaf Bat	
H. lankadiva (Or)	Indian Roundleaf Bat	
H. larvatus (P, Or, A)	Intermediate Roundleaf Bat	[115]
H. lekaguli (Or)	Large Asian Roundleaf Bat	[48]
H. lylei (P, Or)	Shield-faced Roundleaf Bat	
H. macrobullatus (A)	Big-eared Roundleaf Bat	
H. maggietaylorae (A)	Maggie Taylor's Roundleaf Bat	
H. marisae (E)	Aellen's Roundleaf Bat	
H. megalotis (P, E)	Ethiopian Large-eared Roundleaf Bat	
H. muscinus (A)	Fly River Roundleaf Bat	
H. nequam (Or)	Malayan Roundleaf Bat	
H. obscurus (Or)	Philippine Forest Roundleaf Bat	[48]
H. papua (A)	Biak Roundleaf Bat	
H. pomona (P, Or)	Pomona Roundleaf Bat	
H. pratti (P, Or)	Pratt's Roundleaf Bat	
H. pygmaeus (Or)	Philippine Pygmy Roundleaf Bat	[48]
H. ridleyi (Or)	Ridley's Roundleaf Bat	[91]
H. ruber (E)	Noack's Roundleaf Bat	
H. sabanus (Or)	Least Roundleaf Bat	[91]
H. schistaceus (Or)	Split Roundleaf Bat	
H. semoni (A)	Semon's Roundleaf Bat	
H. speoris (Or)	Schneider's Roundleaf Bat	[27]
H. stenotis (A)	Narrow-eared Roundleaf Bat	
H. turpis (P, Or)	Lesser Roundleaf Bat	
H. wollastoni (A)	Wollaston's Roundleaf Bat	
Genus *Paracoelops*		
P. megalotis (Or)	Vietnam Leaf-nosed Bat	

Genus *Rhinonicteris*		
R. *aurantia* (A)	Orange Leaf-nosed Bat	[110]
Genus *Triaenops*	Triple Nose-leaf Bats	[86]
T. *furculus* (Oc, E)	Trouessart's Trident Bat	[16]
T. *persicus* (P, E)	Persian Trident Bat	[16]
Family Noctilionidae	Bulldog Bats	[26]
Genus *Noctilio*		
N. *albiventris* (Nea, Neo)	Lesser Bulldog Bat	[26]
N. *leporinus* (Nea, Neo)	Greater Bulldog Bat	[26]
Family Mormoopidae	Leaf-chinned Bats	[98]
Genus *Mormoops*	Ghost-faced Bats	[38]
M. *blainvillii* (Neo)	Antillean Ghost-faced Bat	[38]
M. *megalophylla* (Nea, Neo)	Ghost-faced Bat	[114]
Genus *Pteronotus*	Mustached Bats	[86]
P. *davyi* (Nea, Neo)	Davy's Naked-backed Bat	[98]
P. *gymnonotus* (Nea, Neo)	Big Naked-backed Bat	[98]
P. *macleayii* (Neo)	MacLeay's Mustached Bat	[38]
P. *parnellii* (Nea, Neo)	Parnell's Mustached Bat	[38]
P. *personatus* (Nea, Neo)	Wagner's Mustached Bat	[38]
P. *quadridens* (Neo)	Sooty Mustached Bat	[38]
Family Phyllostomidae	American Leaf-nosed Bats	[86]
Subfamily Phyllostominae		
Genus *Chrotopterus*		
C. *auritus* (Nea, Neo)	Big-eared Woolly Bat	
Genus *Lonchorhina*	Sword-nosed Bats	[86]
L. *aurita* (Nea, Neo)	Tomes's Sword-nosed Bat	
L. *fernandezi* (Neo)	Fernandez's Sword-nosed Bat	
L. *marinkellei* (Neo)	Marinkelle's Sword-nosed Bat	[27]
L. *orinocensis* (Neo)	Orinoco Sword-nosed Bat	
Genus *Macrophyllum*		
M. *macrophyllum* (Nea, Neo)	Long-legged Bat	[86]
Genus *Macrotus*	Leaf-nosed Bats	[38]
M. *californicus* (Nea)	California Leaf-nosed Bat	[114]
M. *waterhousii* (Nea, Neo)	Waterhouse's Leaf-nosed Bat	[38]
Genus *Micronycteris*	Little Big-eared Bats	[86]
M. *behnii* (Neo)	Behni's Big-eared Bat	
M. *brachyotis* (Nea, Neo)	Yellow-throated Big-eared Bat	[34]

M. daviesi (Neo)	Davies's Big-eared Bat	[38]
M. hirsuta (Neo)	Hairy Big-eared Bat	[34]
M. megalotis (Nea, Neo)	Little Big-eared Bat	[34]
M. minuta (Neo)	White-bellied Big-eared Bat	[34]
M. nicefori (Neo)	Niceforo's Big-eared Bat	[34]
M. pusilla (Neo)	Least Big-eared Bat	
M. schmidtorum (Nea, Neo)	Schmidts's Big-eared Bat	[38]
M. sylvestris (Nea, Neo)	Tri-colored Big-eared Bat	[34]
Genus *Mimon*	Hairy-nosed Bats	[26]
M. bennettii (Nea, Neo)	Golden Bat	[98]
M. crenulatum (Nea, Neo)	Striped Hairy-nosed Bat	[98]
Genus *Phylloderma*		
P. stenops (Nea, Neo)	Pale-faced Bat	[98]
Genus *Phyllostomus*	Spear-nosed Bats	[86]
P. discolor (Nea, Neo)	Pale Spear-nosed Bat	[98]
P. elongatus (Neo)	Lesser Spear-nosed Bat	
P. hastatus (Neo)	Greater Spear-nosed Bat	[98]
P. latifolius (Neo)	Guianan Spear-nosed Bat	
Genus *Tonatia*	Round-eared Bats	[86]
T. bidens (Neo)	Greater Round-eared Bat	[34]
T. brasiliense (Nea, Neo)	Pygmy Round-eared Bat	[38]
T. carrikeri (Neo)	Carriker's Round-eared Bat	
T. evotis (Nea, Neo)	Davis's Round-eared Bat	[98]
T. schulzi (Neo)	Schultz's Round-eared Bat	
T. silvicola (Neo)	White-throated Round-eared Bat	[98]
Genus *Trachops*		
T. cirrhosus (Nea, Neo)	Fringe-lipped Bat	[26]
Genus *Vampyrum*		
V. spectrum (Nea, Neo)	Spectral Bat	[98]
Subfamily Lonchophyllinae		
Genus *Lionycteris*		
L. spurrelli (Neo)	Chestnut Long-tongued Bat	[98]
Genus *Lonchophylla*	Nectar Bats	[98]
L. bokermanni (Neo)	Bokermann's Nectar Bat	
L. dekeyseri (Neo)	Dekeyser's Nectar Bat	
L. handleyi (Neo)	Handley's Nectar Bat	
L. hesperia (Neo)	Western Nectar Bat	
L. mordax (Neo)	Goldman's Nectar Bat	[98]

L. robusta (Neo)	Orange Nectar Bat	[98]
L. thomasi (Neo)	Thomas's Nectar Bat	[98]
Genus *Platalina*		
P. genovensium (Neo)	Long-snouted Bat	
Subfamily Brachyphyllinae		
Genus *Brachyphylla*	West Indian Fruit-eating Bats	[109]
B. cavernarum (Neo)	Antillean Fruit-eating Bat	[38]
B. nana (Neo)	Cuban Fruit-eating Bat	[16]
Subfamily Phyllonycterinae		
Genus *Erophylla*		
E. sezekorni (Neo)	Buffy Flower Bat	[38]
Genus *Phyllonycteris*	Smooth-toothed Flower Bats	[109]
P. aphylla (Neo)	Jamaican Flower Bat	[38]
P. poeyi (Neo)	Cuban Flower Bat	[38]
Subfamily Glossophaginae		
Genus *Anoura*	Tailless Bats	[34]
A. caudifer (Neo)	Tailed Tailless Bat	
A. cultrata (Neo)	Handley's Tailless Bat	[38]
A. geoffroyi (Nea, Neo)	Geoffroy's Tailless Bat	[38]
A. latidens (Neo)	Broad-toothed Tailless Bat	
Genus *Choeroniscus*	Long-tailed Bats	[38]
C. godmani (Nea, Neo)	Godman's Long-tailed Bat	
C. intermedius (Neo)	Intermediate Long-tailed Bat	
C. minor (Neo)	Lesser Long-tailed Bat	
C. periosus (Neo)	Greater Long-tailed Bat	
Genus *Choeronycteris*		
C. mexicana (Nea, Neo)	Mexican Long-tongued Bat	[38]
Genus *Glossophaga*	Long-tongued Bats	[38]
G. commissarisi (Nea, Neo)	Commissaris's Long-tongued Bat	[38]
G. leachii (Nea, Neo)	Gray Long-tongued Bat	[98]
G. longirostris (Neo)	Miller's Long-tongued Bat	[38]
G. morenoi (Nea)	Western Long-tongued Bat	[98]
G. soricina (Nea, Neo)	Pallas's Long-tongued Bat	[38]
Genus *Hylonycteris*		
H. underwoodi (Nea, Neo)	Underwood's Long-tongued Bat	[26]

Genus *Leptonycteris*	Long-nosed Bats	[38]
L. *curasoae* (Nea, Neo)	Southern Long-nosed Bat	[114]
L. *nivalis* (Nea, Neo)	Mexican Long-nosed Bat	[114]
Genus *Lichonycteris*		
L. *obscura* (Neo)	Dark Long-tongued Bat	[98]
Genus *Monophyllus*	Single Leaf Bats	
M. *plethodon* (Neo)	Insular Single Leaf Bat	
M. *redmani* (Neo)	Leach's Single Leaf Bat	
Genus *Musonycteris*		
M. *harrisoni* (Nea)	Banana Bat	[86]
Genus *Scleronycteris*		
S. *ega* (Neo)	Ega Long-tongued Bat	[26]
Subfamily Carolliinae		
Genus *Carollia*	Short-tailed Bats	[38]
C. *brevicauda* (Nea, Neo)	Silky Short-tailed Bat	[98]
C. *castanea* (Neo)	Chestnut Short-tailed Bat	[98]
C. *perspicillata* (Nea, Neo)	Seba's Short-tailed Bat	[98]
C. *subrufa* (Nea, Neo)	Gray Short-tailed Bat	[98]
Genus *Rhinophylla*	Little Fruit Bats	[26]
R. *alethina* (Neo)	Hairy Little Fruit Bat	
R. *fischerae* (Neo)	Fischer's Little Fruit Bat	
R. *pumilio* (Neo)	Dwarf Little Fruit Bat	
Subfamily Stenodermatinae		
Genus *Ametrida*		
A. *centurio* (Neo)	Little White-shouldered Bat	[98]
Genus *Ardops*		
A. *nichollsi* (Neo)	Tree Bat	[86]
Genus *Ariteus*		
A. *flavescens* (Neo)	Jamaican Fig-eating Bat	[86]
Genus *Artibeus*	Fruit-eating Bats	[38]
A. *amplus* (Neo)	Large Fruit-eating Bat	
A. *anderseni* (Neo)	Andersen's Fruit-eating Bat	
A. *aztecus* (Nea, Neo)	Aztec Fruit-eating Bat	[98]
A. *cinereus* (Neo)	Gervais's Fruit-eating Bat	
A. *concolor* (Neo)	Brown Fruit-eating Bat	
A. *fimbriatus* (Neo)	Fringed Fruit-eating Bat	
A. *fraterculus* (Neo)	Fraternal Fruit-eating Bat	
A. *glaucus* (Nea, Neo)	Silver Fruit-eating Bat	
A. *hartii* (Nea, Neo)	Velvety Fruit-eating Bat	[98]
A. *hirsutus* (Nea)	Hairy Fruit-eating Bat	

A. inopinatus (Neo)	Honduran Fruit-eating Bat	[98]
A. jamaicensis (Nea, Neo)	Jamaican Fruit-eating Bat	[98]
A. lituratus (Nea, Neo)	Great Fruit-eating Bat	[98]
A. obscurus (Neo)	Dark Fruit-eating Bat	
A. phaeotis (Nea, Neo)	Pygmy Fruit-eating Bat	[98]
A. planirostris (Neo)	Flat-faced Fruit-eating Bat	
A. toltecus (Nea, Neo)	Toltec Fruit-eating Bat	[98]
Genus *Centurio*		
C. senex (Nea, Neo)	Wrinkle-faced Bat	[26]
Genus *Chiroderma*	Big-eyed Bats	[26]
C. doriae (Neo)	Brazilian Big-eyed Bat	
C. improvisum (Neo)	Guadeloupe Big-eyed Bat	
C. salvini (Nea, Neo)	Salvin's Big-eyed Bat	[98]
C. trinitatum (Neo)	Little Big-eyed Bat	[98]
C. villosum (Nea, Neo)	Hairy Big-eyed Bat	[98]
Genus *Ectophylla*		
E. alba (Neo)	White Bat	[86]
Genus *Mesophylla*		
M. macconnelli (Neo)	MacConnell's Bat	[26]
Genus *Phyllops*		
P. falcatus (Neo)	Cuban Fig-eating Bat	[38]
Genus *Platyrrhinus*	Broad-nosed Bats	
P. aurarius (Neo)	Eldorado Broad-nosed Bat	
P. brachycephalus (Neo)	Short-headed Broad-nosed Bat	
P. chocoensis (Neo)	Choco Broad-nosed Bat	
P. dorsalis (Neo)	Thomas's Broad-nosed Bat	[98]
P. helleri (Nea, Neo)	Heller's Broad-nosed Bat	[98]
P. infuscus (Neo)	Buffy Broad-nosed Bat	
P. lineatus (Neo)	White-lined Broad-nosed Bat	
P. recifinus (Neo)	Recife Broad-nosed Bat	
P. umbratus (Neo)	Shadowy Broad-nosed Bat	
P. vittatus (Neo)	Greater Broad-nosed Bat	[98]
Genus *Pygoderma*		
P. bilabiatum (Neo)	Ipanema Bat	[26]
Genus *Sphaeronycteris*		
S. toxophyllum (Neo)	Visored Bat	[26]
Genus *Stenoderma*		
S. rufum (Neo)	Red Fruit Bat	[86]

Genus *Sturnira* — Yellow-shouldered Bats [86]

  S. aratathomasi (Neo) — Aratathomas's Yellow-shouldered Bat

  S. bidens (Neo) — Bidentate Yellow-shouldered Bat

  S. bogotensis (Neo) — Bogota Yellow-shouldered Bat

  S. erythromos (Neo) — Hairy Yellow-shouldered Bat

  S. lilium (Nea, Neo) — Little Yellow-shouldered Bat [98]

  S. ludovici (Nea, Neo) — Highland Yellow-shouldered Bat [98]

  S. luisi (Neo) — Luis's Yellow-shouldered Bat [98]

  S. magna (Neo) — Greater Yellow-shouldered Bat

  S. mordax (Neo) — Talamancan Yellow-shouldered Bat [98]

  S. nana (Neo) — Lesser Yellow-shouldered Bat

  S. thomasi (Neo) — Thomas's Yellow-shouldered Bat

  S. tildae (Neo) — Tilda Yellow-shouldered Bat

Genus *Uroderma* — Tent-making Bats [26]

  U. bilobatum (Nea, Neo) — Tent-making Bat [38]

  U. magnirostrum (Nea, Neo) — Brown Tent-making Bat [98]

Genus *Vampyressa* — Yellow-eared Bats [86]

  V. bidens (Neo) — Bidentate Yellow-eared Bat

  V. brocki (Neo) — Brock's Yellow-eared Bat

  V. melissa (Neo) — Melissa's Yellow-eared Bat

  V. nymphaea (Neo) — Striped Yellow-eared Bat [98]

  V. pusilla (Nea, Neo) — Little Yellow-eared Bat [98]

Genus *Vampyrodes*

  V. caraccioli (Nea, Neo) — Great Stripe-faced Bat [86]

Subfamily Desmodontinae

Genus *Desmodus*

  D. rotundus (Nea, Neo) — Vampire Bat [38]

Genus *Diaemus*

D. youngi (Nea, Neo)   White-winged Vampire Bat   [86]

Genus *Diphylla*

D. ecaudata (Nea, Neo)   Hairy-legged Vampire Bat   [86]

Family Natalidae   Funnel-eared Bats   [86]

Genus *Natalus*

N. lepidus (Neo)   Gervais's Funnel-eared Bat   [38]

N. micropus (Neo)   Cuban Funnel-eared Bat   [38]

N. stramineus (Nea, Neo)   Mexican Funnel-eared Bat   [98]

N. tumidifrons (Neo)   Bahaman Funnel-eared Bat

N. tumidirostris (Neo)   Trinidadian Funnel-eared

    Bat   [34]

Family Furipteridae   Thumbless Bats   [86]

Genus *Amorphochilus*

A. schnablii (Neo)   Smoky Bat   [109]

Genus *Furipterus*

F. horrens (Neo)   Thumbless Bat   [98]

Family Thyropteridae   Disk-winged Bats   [86]

Genus *Thyroptera*

T. discifera (Neo)   Peter's Disk-winged Bat   [98]

T. tricolor (Nea, Neo)   Spix's Disk-winged Bat   [98]

Family Myzopodidae

Genus *Myzopoda*

M. aurita (E)   Sucker-footed Bat   [16]

Family Vespertilionidae   Vesper Bats   [16]

Subfamily Kerivoulinae

Genus *Kerivoula*   Woolly Bats   [86]

K. aerosa (E)   Dubious Trumpet-eared

    Bat   [16]

K. africana (E)   Tanzanian Woolly Bat   [16]

K. agnella (A)   St. Aignan's Trumpet-eared

    Bat   [29]

K. argentata (E)   Damara Woolly Bat   [16]

K. atrox (Or)   Groove-toothed Bat   [16]

K. cuprosa (E)   Copper Woolly Bat   [16]

K. eriophora (E)   Ethiopian Woolly Bat

K. flora (Or, A)   Flores Woolly Bat

K. hardwickei (P, Or, A)   Hardwicke's Woolly Bat   [91]

K. intermedia (Or)   Small Woolly Bat   [91]

K. jagori (Or, A)   Peters's Trumpet-eared Bat   [16]

K. lanosa (E)	Lesser Woolly Bat	[16]
K. minuta (Or)	Least Woolly Bat	[91]
K. muscina (A)	Fly River Trumpet-eared Bat	[16]
K. myrella (A)	Bismarck's Trumpet-eared Bat	[16]
K. papillosa (Or, A)	Papillose Woolly Bat	[91]
K. papuensis (A)	Golden-tipped Bat	[110]
K. pellucida (Or)	Clear-winged Woolly Bat	[91]
K. phalaena (E)	Spurrell's Woolly Bat	[16]
K. picta (P, Or, A)	Painted Bat	[16]
K. smithi (E)	Smith's Woolly Bat	[16]
K. whiteheadi (Or)	Whitehead's Woolly Bat	[91]
Subfamily Vespertilioninae		
Genus *Antrozous*	Pallid Bats	[86]
A. dubiaquercus (Nea, Neo)	Van Gelder's Bat	[98]
A. pallidus (Nea, Neo)	Pallid Bat	[114]
Genus *Barbastella*	Barbastelles	[16]
B. barbastellus (P)	Western Barbastelle	[16]
B. leucomelas (P, E, Or)	Eastern Barbastelle	[16]
Genus *Chalinolobus*	Wattled Bats	[86]
C. alboguttatus (E)	Allen's Striped Bat	[16]
C. argentatus (E)	Silvered Bat	[16]
C. beatrix (E)	Beatrix's Bat	[16]
C. dwyeri (A)	Large-eared Pied Bat	[110]
C. egeria (E)	Bibundi Bat	[103]
C. gleni (E)	Glen's Wattled Bat	
C. gouldii (A)	Gould's Wattled Bat	[110]
C. kenyacola (E)	Kenyan Wattled Bat	
C. morio (A)	Chocolate Wattled Bat	[110]
C. nigrogriseus (A)	Hoary Wattled Bat	[110]
C. picatus (A)	Little Pied Bat	[110]
C. poensis (E)	Abo Bat	[16]
C. superbus (E)	Pied Bat	[103]
C. tuberculatus (A)	Long-tailed Wattled Bat	
C. variegatus (E)	Butterfly Bat	[16]
Genus *Eptesicus*	Serotines	[16]
E. baverstocki (A)	Inland Forest Bat	[110]
E. bobrinskii (P)	Bobrinski's Serotine	
E. bottae (P)	Botta's Serotine	[16]

E. brasiliensis (Nea, Neo)	Brazilian Brown Bat	[6]
E. brunneus (E)	Dark-brown Serotine	[16]
E. capensis (E)	Cape Serotine	[16]
E. demissus (Or)	Surat Serotine	[16]
E. diminutus (Neo)	Diminutive Serotine	
E. douglasorum (A)	Yellow-lipped Bat	[16]
E. flavescens (E)	Yellow Serotine	[16]
E. floweri (E)	Horn-skinned Bat	[16]
E. furinalis (Nea, Neo)	Argentine Brown Bat	[6]
E. fuscus (Nea, Neo)	Big Brown Bat	[114]
E. guadeloupensis (Neo)	Guadeloupe Big Brown Bat	[38]
E. guineensis (E)	Tiny Serotine	[16]
E. hottentotus (E)	Long-tailed House Bat	[16]
E. innoxius (Neo)	Harmless Serotine	
E. kobayashii (P)	Kobayashi's Serotine	
E. melckorum (E)	Melck's House Bat	[16]
E. nasutus (P)	Sind Bat	[16]
E. nilssoni (P, Or)	Northern Bat	[16]
E. pachyotis (Or)	Thick-eared Bat	[16]
E. platyops (E)	Lagos Serotine	[16]
E. pumilus (A)	Eastern Forest Bat	[110]
E. regulus (A)	Southern Forest Bat	[110]
E. rendalli (E)	Rendall's Serotine	[16]
E. sagittula (Oc, A)	Large Forest Bat	[16]
E. serotinus (P, Or)	Serotine	[16]
E. somalicus (E)	Somali Serotine	[16]
E. tatei (Or)	Sombre Bat	[8]
E. tenuipinnis (E)	White-winged Serotine	[16]
E. vulturnus (A)	Little Forest Bat	[16]
Genus *Euderma*		
E. maculatum (Nea)	Spotted Bat	[114]
Genus *Eudiscopus*		
E. denticulus (Or)	Disk-footed Bat	[16]
Genus *Glischropus*	Thick-thumbed Bats	[86]
G. javanus (Or)	Javan Thick-thumbed Bat	
G. tylopus (Or, A)	Common Thick-thumbed Bat	[29]
Genus *Hesperoptenus*	False Serotines	
H. blanfordi (Or)	Blanford's Bat	[16]
H. doriae (Or)	False Serotine Bat	[16]

H. gaskelli (A)	Gaskell's False Serotine	
H. tickelli (P, Or)	Tickell's Bat	[16]
H. tomesi (Or)	Large False Serotine	[109]
Genus *Histiotus*	Big-eared Brown Bats	[86]
H. alienus (Neo)	Strange Big-eared Brown Bat	
H. macrotus (Neo)	Big-eared Brown Bat	[7]
H. montanus (Neo)	Small Big-eared Brown Bat	[6]
H. velatus (Neo)	Tropical Big-eared Brown Bat	[6]
Genus *Ia*		
I. io (P, Or)	Great Evening Bat	[16]
Genus *Idionycteris*		
I. phyllotis (Nea)	Allen's Big-eared Bat	[114]
Genus *Laephotis*	African Long-eared Bats	[86]
L. angolensis (E)	Angolan Long-eared Bat	
L. botswanae (E)	Botswanan Long-eared Bat	[16]
L. namibensis (E)	Namib Long-eared Bat	[16]
L. wintoni (E)	De Winton's Long-eared Bat	[16]
Genus *Lasionycteris*		
L. noctivagans (Nea, Neo)	Silver-haired Bat	[114]
Genus *Lasiurus*	Hairy-tailed Bats	[86]
L. borealis (Nea, Neo)	Red Bat	[114]
L. castaneus (Neo)	Tacarcuna Bat	[38]
L. cinereus (Oc, Nea, Neo)	Hoary Bat	[114]
L. ega (Nea, Neo)	Southern Yellow Bat	[114]
L. egregius (Neo)	Big Red Bat	[24]
L. intermedius (Nea, Neo)	Northern Yellow Bat	[114]
L. seminolus (Nea)	Seminole Bat	[114]
Genus *Mimetillus*		
M. moloneyi (E)	Moloney's Flat-headed Bat	[16]
Genus *Myotis*	Little Brown Bats	[16]
M. abei (P)	Sakhalin Myotis	
M. adversus (Or, A)	Large-footed Bat	[16]
M. aelleni (Neo)	Southern Myotis	[7]
M. albescens (Nea, Neo)	Silver-tipped Myotis	[7]
M. altarium (P, Or)	Szechwan Myotis	
M. annectans (Or)	Hairy-faced Bat	[16]
M. atacamensis (Neo)	Atacama Myotis	

M. auriculus (Nea, Neo)	Southwestern Myotis	[114]
M. australis (A)	Australian Myotis	
M. austroriparius (Nea)	Southeastern Myotis	[114]
M. bechsteini (P)	Bechstein's Bat	[16]
M. blythii (P, Or)	Lesser Mouse-eared Bat	[16]
M. bocagei (P, E)	Rufous Mouse-eared Bat	[16]
M. bombinus (P)	Far Eastern Myotis	
M. brandti (P)	Brandt's Bat	[16]
M. californicus (Nea, Neo)	California Myotis	[114]
M. capaccinii (P)	Long-fingered Bat	[16]
M. chiloensis (Neo)	Chilean Myotis	[6]
M. chinensis (P, Or)	Large Myotis	[16]
M. cobanensis (Neo)	Guatemalan Myotis	[98]
M. dasycneme (P)	Pond Bat	[16]
M. daubentoni (P, Or)	Daubenton's Bat	[16]
M. dominicensis (Neo)	Dominican Myotis	[16]
M. elegans (Nea, Neo)	Elegant Myotis	[38]
M. emarginatus (P)	Geoffroy's Bat	[16]
M. evotis (Nea)	Long-eared Myotis	[114]
M. findleyi (Nea)	Findley's Myotis	[16]
M. formosus (P, Or, A)	Hodgson's Bat	[16]
M. fortidens (Nea, Neo)	Cinnamon Myotis	[38]
M. frater (P)	Fraternal Myotis	
M. goudoti (E)	Malagasy Mouse-eared Bat	[16]
M. grisescens (Nea)	Gray Myotis	[114]
M. hasseltii (Or)	Lesser Large-footed Bat	[16]
M. horsfieldii (P, Or, A)	Horsfield's Bat	[8]
M. hosonoi (P)	Hosono's Myotis	
M. ikonnikovi (P)	Ikonnikov's Bat	[109]
M. insularum (Oc)	Insular Myotis	
M. keaysi (Nea, Neo)	Hairy-legged Myotis	[7]
M. keenii (Nea)	Keen's Myotis	[114]
M. leibii (Nea)	Eastern Small-footed Myotis	[114]
M. lesueuri (E)	Lesueur's Hairy Bat	[16]
M. levis (Neo)	Yellowish Myotis	[6]
M. longipes (P, Or)	Kashmir Cave Bat	[8]
M. lucifugus (Nea)	Little Brown Bat	[114]
M. macrodactylus (P)	Big-footed Myotis	
M. macrotarsus (Or)	Pallid Large-footed Myotis	[27]

M. martiniquensis (Neo)	Schwartz's Myotis	[16]
M. milleri (Nea)	Miller's Myotis	[38]
M. montivagus (P, Or)	Burmese Whiskered Bat	[16]
M. morrisi (E)	Morris's Bat	
M. muricola (P, Or, A)	Whiskered Myotis	[91]
M. myotis (P)	Mouse-eared Bat	[19]
M. mystacinus (P, Or)	Whiskered Bat	[16]
M. nattereri (P)	Natterer's Bat	[16]
M. nesopolus (Neo)	Curacao Myotis	
M. nigricans (Nea, Neo)	Black Myotis	[7]
M. oreias (Or)	Singapore Whiskered Bat	[16]
M. oxyotus (Neo)	Montane Myotis	[38]
M. ozensis (P)	Honshu Myotis	
M. peninsularis (Nea)	Peninsular Myotis	[38]
M. pequinius (P)	Peking Myotis	
M. planiceps (Nea)	Flat-headed Myotis	[38]
M. pruinosus (P)	Frosted Myotis	
M. ricketti (P)	Rickett's Big-footed Bat	[16]
M. ridleyi (Or)	Ridley's Bat	[16]
M. riparius (Neo)	Riparian Myotis	[98]
M. rosseti (Or)	Thick-thumbed Myotis	[16]
M. ruber (Neo)	Red Myotis	[6]
M. schaubi (P)	Schaub's Myotis	
M. scotti (E)	Scott's Mouse-eared Bat	[16]
M. seabrai (E)	Angolan Hairy Bat	[16]
M. sicarius (Or)	Mandelli's Mouse-eared Bat	[8]
M. siligorensis (P, Or)	Himalayan Whiskered Bat	[16]
M. simus (Neo)	Velvety Myotis	[6]
M. sodalis (Nea)	Indiana Bat	[114]
M. stalkeri (A)	Kei Myotis	[16]
M. thysanodes (Nea)	Fringed Myotis	[114]
M. tricolor (E)	Cape Hairy Bat	[16]
M. velifer (Nea, Neo)	Cave Myotis	[114]
M. vivesi (Nea)	Fish-eating Bat	[38]
M. volans (Nea)	Long-legged Myotis	[114]
M. welwitschii (E)	Welwitch's Bat	[44]
M. yesoensis (P)	Yoshiyuki's Myotis	
M. yumanensis (Nea)	Yuma Myotis	[114]
Genus *Nyctalus*	Noctule Bats	[86]
N. aviator (P)	Birdlike Noctule	

N. azoreum (Oc)	Azores Noctule	
N. lasiopterus (P)	Giant Noctule	[16]
N. leisleri (Oc, P)	Lesser Noctule	[16]
N. montanus (P, Or)	Mountain Noctule	[8]
N. noctula (P, Or)	Noctule	[16]
Genus *Nycticeius*	Broad-nosed Bats	
N. balstoni (A)	Western Broad-nosed Bat	[16]
N. greyii (A)	Little Broad-nosed Bat	[16]
N. humeralis (Nea, Neo)	Evening Bat	[114]
N. rueppellii (A)	Rüppell's Broad-nosed Bat	[16]
N. sanborni (A)	Northern Broad-nosed Bat	[110]
N. schlieffeni (P, E)	Schlieffen's Bat	[16]
Genus *Nyctophilus*	Long-eared Bats	[110]
N. arnhemensis (A)	Northern Long-eared Bat	[110]
N. geoffroyi (A)	Lesser Long-eared Bat	[110]
N. gouldi (A)	Gould's Long-eared Bat	[110]
N. heran (A)	Sunda Long-eared Bat	
N. microdon (A)	Small-toothed Long-eared Bat	[16]
N. microtis (A)	New Guinea Long-eared Bat	[29]
N. timoriensis (A)	Greater Long-eared Bat	[110]
N. walkeri (A)	Pygmy Long-eared Bat	[110]
Genus *Otonycteris*		
O. hemprichi (P, E, Or)	Desert Long-eared Bat	[86]
Genus *Pharotis*		
P. imogene (A)	New Guinea Big-eared Bat	[86]
Genus *Philetor*		
P. brachypterus (Or, A)	Rohu's Bat	[29]
Genus *Pipistrellus*	Pipistrelles	[16]
P. aegyptius (P, E)	Egyptian Pipistrelle	
P. aero (E)	Mt. Gargues Pipistrelle	[109]
P. affinis (P, Or)	Chocolate Pipistrelle	[8]
P. anchietai (E)	Anchieta's Pipistrelle	[16]
P. anthonyi (Or)	Anthony's Pipistrelle	
P. arabicus (P)	Arabian Pipistrelle	[43]
P. ariel (P, E)	Desert Pipistrelle	[16]
P. babu (P, Or)	Himalayan Pipistrelle	[16]
P. bodenheimeri (P)	Bodenheimer's Pipistrelle	[16]
P. cadornae (Or)	Cadorna's Pipistrelle	[16]

P. ceylonicus (P, Or)	Kelaart's Pipistrelle	[16]
P. circumdatus (P, Or)	Black Gilded Pipistrelle	[17]
P. coromandra (P, Or)	Indian Pipistrelle	[16]
P. crassulus (E)	Broad-headed Pipistrelle	[109]
P. cuprosus (Or)	Coppery Pipistrelle	[91]
P. dormeri (P, Or)	Dormer's Pipistrelle	
P. eisentrauti (E)	Eisentraut's Pipistrelle	[16]
P. endoi (P)	Endo's Pipistrelle	
P. hesperus (Nea)	Western Pipistrelle	[114]
P. imbricatus (Or, A)	Brown Pipistrelle	[16]
P. inexspectatus (E)	Aellen's Pipistrelle	[16]
P. javanicus (P, Or, A)	Javan Pipistrelle	[16]
P. joffrei (Or)	Joffre's Pipistrelle	[16]
P. kitcheneri (Or)	Red-brown Pipistrelle	[91]
P. kuhlii (P, E)	Kuhl's Pipistrelle	[16]
P. lophurus (Or)	Burma Pipistrelle	
P. macrotis (Or)	Big-eared Pipistrelle	
P. maderensis (P)	Madeira Pipistrelle	[16]
P. mimus (P, Or)	Indian Pygmy Pipistrelle	[16]
P. minahassae (A)	Minahassa Pipistrelle	[16]
P. mordax (Or)	Pungent Pipistrelle	
P. musciculus (E)	Mouselike Pipistrelle	
P. nanulus (E)	Tiny Pipistrelle	[16]
P. nanus (E)	Banana Pipistrelle	
P. nathusii (P)	Nathusius's Pipistrelle	[16]
P. paterculus (P, Or)	Mount Popa Pipistrelle	[8]
P. peguensis (Or)	Pegu Pipistrelle	
P. permixtus (E)	Dar-es-salaam Pipistrelle	[16]
P. petersi (Or, A)	Peters's Pipistrelle	[16]
P. pipistrellus (P, Or)	Common Pipistrelle	[16]
P. pulveratus (P, Or)	Chinese Pipistrelle	[16]
P. rueppelli (P, E)	Rüppel's Pipistrelle	[103]
P. rusticus (E)	Rusty Pipistrelle	[27]
P. savii (P, Or)	Savi's Pipistrelle	[16]
P. societatis (Or)	Social Pipistrelle	
P. stenopterus (Or)	Narrow-winged Pipistrelle	[91]
P. sturdeei (P)	Sturdee's Pipistrelle	
P. subflavus (Nea, Neo)	Eastern Pipistrelle	[114]
P. tasmaniensis (A)	Eastern False Pipistrelle	[110]
P. tenuis (P, Or, A)	Least Pipistrelle	[17]

Genus *Plecotus*	Big-eared Bats	[38]
P. *auritus* (P, Or)	Brown Big-eared Bat	
P. *austriacus* (Oc, P, E)	Gray Big-eared Bat	
P. *mexicanus* (Nea)	Mexican Big-eared Bat	[16]
P. *rafinesquii* (Nea)	Rafinesque's Big-eared Bat	[16]
P. *taivanus* (P)	Taiwan Big-eared Bat	
P. *teneriffae* (P)	Canary Big-eared Bat	
P. *townsendii* (Nea)	Townsend's Big-eared Bat	[16]
Genus *Rhogeessa*	Little Yellow Bats	[86]
R. *alleni* (Nea)	Allen's Yellow Bat	[38]
R. *genowaysi* (Nea)	Genoways's Yellow Bat	
R. *gracilis* (Nea)	Slender Yellow Bat	[38]
R. *minutilla* (Neo)	Tiny Yellow Bat	
R. *mira* (Nea)	Least Yellow Bat	[38]
R. *parvula* (Nea)	Little Yellow Bat	[38]
R. *tumida* (Nea, Neo)	Black-winged Little Yellow Bat	[26]
Genus *Scotoecus*	House Bats	
S. *albofuscus* (E)	Light-winged Lesser House Bat	[44]
S. *hirundo* (E)	Dark-winged Lesser House Bat	[44]
S. *pallidus* (P, Or)	Desert Yellow Bat	[8]
Genus *Scotomanes*	Harlequin Bats	[86]
S. *emarginatus* (Or)	Emarginate Harlequin Bat	
S. *ornatus* (P, Or)	Harlequin Bat	[16]
Genus *Scotophilus*	Yellow Bats	[86]
S. *borbonicus* (Oc, E)	Lesser Yellow Bat	[27]
S. *celebensis* (A)	Sulawesi Yellow Bat	[16]
S. *dinganii* (E)	African Yellow Bat	
S. *heathi* (P, Or)	Greater Asiatic Yellow Bat	
S. *kuhlii* (P, Or, A)	Lesser Asiatic Yellow Bat	
S. *leucogaster* (P, E)	White-bellied Yellow Bat	
S. *nigrita* (E)	Schreber's Yellow Bat	
S. *nux* (E)	Nut-colored Yellow Bat	
S. *robustus* (E)	Robust Yellow Bat	
S. *viridis* (E)	Greenish Yellow Bat	
Genus *Tylonycteris*	Bamboo Bats	[86]
T. *pachypus* (P, Or)	Lesser Bamboo Bat	[27]
T. *robustula* (P, Or, A)	Greater Bamboo Bat	[27]

Genus *Vespertilio*   Particolored Bats   [86]
    V. murinus (P, Or)   Particolored Bat   [16]
    V. superans (P)   Asian Particolored Bat
Subfamily Murininae
Genus *Harpiocephalus*
    H. harpia (P, Or, A)   Hairy-winged Bat   [16]
Genus *Murina*   Tube-nosed Insectivorous
                                      Bats   [86]
    M. aenea (Or)   Bronze Tube-nosed Bat   [16]
    M. aurata (P, Or)   Little Tube-nosed Bat   [16]
    M. cyclotis (P, Or, A)   Round-eared Tube-nosed Bat [17]
    M. florium (A)   Flores Tube-nosed Bat   [16]
    M. fusca (P)   Dusky Tube-nosed Bat
    M. grisea (Or)   Peters's Tube-nosed Bat   [16]
    M. huttoni (P, Or)   Hutton's Tube-nosed Bat   [16]
    M. leucogaster (P)   Greater Tube-nosed Bat   [16]
    M. puta (P)   Taiwan Tube-nosed Bat
    M. rozendaali (Or)   Gilded Tube-nosed Bat   [91]
    M. silvatica (P)   Forest Tube-nosed Bat
    M. suilla (Or)   Brown Tube-nosed Bat   [16]
    M. tenebrosa (P)   Gloomy Tube-nosed Bat
    M. tubinaris (P, Or)   Scully's Tube-nosed Bat   [8]
    M. ussuriensis (P)   Ussuri Tube-nosed Bat
Subfamily Miniopterinae
Genus *Miniopterus*   Bent-winged Bats   [86]
    M. australis (Or, A)   Little Long-fingered Bat   [16]
    M. fraterculus (E)   Lesser Long-fingered Bat   [16]
    M. fuscus (P, Or, A)   Southeast Asian Long-
                                      fingered Bat   [109]
    M. inflatus (E)   Greater Long-fingered Bat [16]
    M. magnater (P, Or, A)   Western Bent-winged Bat   [28]
    M. minor (E)   Least Long-fingered Bat   [16]
    M. pusillus (Or, A)   Small Bent-winged Bat   [91]
    M. robustior (A)   Loyalty Bent-winged Bat   [29]
    M. schreibersi (P, E, Or, A) Schreibers's Long-fingered
                                      Bat   [16]
    M. tristis (Or, A)   Great Bent-winged Bat   [29]
Subfamily Tomopeatinae
Genus *Tomopeas*
    T. ravus (Neo)   Blunt-eared Bat

Family Mystacinidae	New Zealand Short-tailed Bats	[16]
Genus *Mystacina*		
M. *robusta* (A)	New Zealand Greater Short-tailed Bat	[16]
M. *tuberculata* (A)	New Zealand Lesser Short-tailed Bat	[16]
Family Molossidae	Free-tailed Bats	[16]
Genus *Chaerephon*	Lesser Free-tailed Bats	
C. *aloysiisabaudiae* (E)	Duke of Abruzzi's Free-tailed Bat	[16]
C. *ansorgei* (E)	Ansorge's Free-tailed Bat	[16]
C. *bemmeleni* (E)	Gland-tailed Free-tailed Bat	[16]
C. *bivittata* (E)	Spotted Free-tailed Bat	[16]
C. *chapini* (E)	Chapin's Free-tailed Bat	[16]
C. *gallagheri* (E)	Gallagher's Free-tailed Bat	[16]
C. *jobensis* (Oc, A)	Northern Mastiff Bat	[16]
C. *johorensis* (Or)	Northern Free-tailed Bat	[110]
C. *major* (E)	Lappet-eared Free-tailed Bat	[44]
C. *nigeriae* (P, E)	Nigerian Free-tailed Bat	[16]
C. *plicata* (Oc, P, Or, A)	Wrinkle-lipped Free-tailed Bat	[16]
C. *pumila* (Oc, P, E)	Little Free-tailed Bat	[16]
C. *russata* (E)	Russet Free-tailed Bat	[16]
Genus *Cheiromeles*		
C. *torquatus* (Or, A)	Hairless Bat	[16]
Genus *Eumops*	Bonneted Bats	[86]
E. *auripendulus* (Nea, Neo)	Black Bonneted Bat	[98]
E. *bonariensis* (Nea, Neo)	Dwarf Bonneted Bat	[98]
E. *dabbenei* (Neo)	Big Bonneted Bat	
E. *glaucinus* (Nea, Neo)	Wagner's Bonneted Bat	[98]
E. *hansae* (Nea, Neo)	Sanborn's Bonneted Bat	[98]
E. *maurus* (Neo)	Guianan Bonneted Bat	
E. *perotis* (Nea, Neo)	Western Bonneted Bat	
E. *underwoodi* (Nea, Neo)	Underwood's Bonneted Bat	[98]
Genus *Molossops*	Dog-faced Bats	[26]
M. *abrasus* (Neo)	Cinnamon Dog-faced Bat	[6]
M. *aequatorianus* (Neo)	Equatorial Dog-faced Bat	
M. *greenhalli* (Nea, Neo)	Greenhall's Dog-faced Bat	[98]
M. *mattogrossensis* (Neo)	Mato Grosso Dog-faced Bat	

M. *neglectus* (Neo)	Rufous Dog-faced Bat	[6]
M. *planirostris* (Neo)	Southern Dog-faced Bat	[98]
M. *temminckii* (Neo)	Dwarf Dog-faced Bat	[6]
Genus *Molossus*	Mastiff Bats	[26]
M. *ater* (Nea, Neo)	Black Mastiff Bat	[98]
M. *bondae* (Nea, Neo)	Bonda Mastiff Bat	[98]
M. *molossus* (Nea, Neo)	Pallas's Mastiff Bat	[7]
M. *pretiosus* (Nea, Neo)	Miller's Mastiff Bat	[98]
M. *sinaloae* (Nea, Neo)	Sinaloan Mastiff Bat	[98]
Genus *Mops*	Greater Free-tailed Bats	
M. *brachypterus* (E)	Sierra Leone Free-tailed Bat	[44]
M. *condylurus* (E)	Angolan Free-tailed Bat	[16]
M. *congicus* (E)	Medje Free-tailed Bat	[16]
M. *demonstrator* (E)	Mongalla Free-tailed Bat	[16]
M. *midas* (P, E)	Midas Free-tailed Bat	[16]
M. *mops* (Or)	Malayan Free-tailed Bat	[16]
M. *nanulus* (E)	Dwarf Free-tailed Bat	[16]
M. *niangarae* (E)	Niangara Free-tailed Bat	[16]
M. *niveiventer* (E)	White-bellied Free-tailed Bat	[16]
M. *petersoni* (E)	Peterson's Free-tailed Bat	[16]
M. *sarasinorum* (Or, A)	Sulawesi Free-tailed Bat	
M. *spurrelli* (E)	Spurrell's Free-tailed Bat	[16]
M. *thersites* (E)	Railer Bat	[16]
M. *trevori* (E)	Trevor's Free-tailed Bat	
Genus *Mormopterus*	Little Mastiff Bats	
M. *acetabulosus* (Oc, E)	Natal Free-tailed Bat	[16]
M. *beccarii* (A)	Beccari's Mastiff Bat	[16]
M. *doriae* (Or)	Sumatran Mastiff Bat	
M. *jugularis* (E)	Peters's Wrinkle-lipped Bat	[16]
M. *kalinowskii* (Neo)	Kalinowski's Mastiff Bat	
M. *minutus* (Neo)	Little Goblin Bat	[38]
M. *norfolkensis* (A)	Eastern Little Mastiff Bat	[16]
M. *petrophilus* (E)	Roberts's Flat-headed Bat	[16]
M. *phrudus* (Neo)	Incan Little Mastiff Bat	
M. *planiceps* (A)	Southern Free-tailed Bat	[110]
M. *setiger* (E)	Peters's Flat-headed Bat	[16]
Genus *Myopterus*	African Free-tailed Bats	
M. *daubentonii* (E)	Daubenton's Free-tailed Bat	[16]
M. *whitleyi* (E)	Bini Free-tailed Bat	[16]

Genus *Nyctinomops*	New World Free-tailed Bats	[86]
N. aurispinosus (Nea, Neo)	Peale's Free-tailed Bat	[38]
N. femorosaccus (Nea)	Pocketed Free-tailed Bat	[114]
N. laticaudatus (Nea, Neo)	Broad-eared Bat	[98]
N. macrotis (Nea, Neo)	Big Free-tailed Bat	[114]
Genus *Otomops*	Big-eared Free-tailed Bats	[86]
O. formosus (Or)	Java Mastiff Bat	[16]
O. martiensseni (E)	Large-eared Free-tailed Bat	[16]
O. papuensis (A)	Big-eared Mastiff Bat	[16]
O. secundus (A)	Mantled Mastiff Bat	[28]
O. wroughtoni (Or)	Wroughton's Free-tailed Bat	[16]
Genus *Promops*	Crested Mastiff Bats	[26]
P. centralis (Nea, Neo)	Big Crested Mastiff Bat	[98]
P. nasutus (Neo)	Brown Mastiff Bat	[6]
Genus *Tadarida*	Free-tailed Bats	[86]
T. aegyptiaca (P, E, Or)	Egyptian Free-tailed Bat	[16]
T. australis (A)	White-striped Free-tailed Bat	[110]
T. brasiliensis (Nea, Neo)	Brazilian Free-tailed Bat	[114]
T. espiritosantensis (Neo)	Espirito Santo Free-tailed Bat	
T. fulminans (E)	Madagascan Large Free-tailed Bat	[16]
T. lobata (E)	Kenyan Big-eared Free-tailed Bat	[16]
T. teniotis (P)	European Free-tailed Bat	[16]
T. ventralis (E)	African Giant Free-tailed Bat	[16]

ORDER PRIMATES

Family Cheirogaleidae

Subfamily Cheirogalinae
Genus *Allocebus*

Genus *Cheirogaleus*

	Primates	[16]
	Dwarf Lemurs and Mouse Lemurs	[16]
A. trichotis (E)	Hairy-eared Dwarf Lemur	[42]
	Dwarf Lemurs	[16]
C. major (E)	Greater Dwarf Lemur	[42]
C. medius (E)	Fat-tailed Dwarf Lemur	[42]

Genus *Microcebus*	Mouse Lemurs	[16]
M. coquereli (E)	Coquerel's Mouse Lemur	[42]
M. murinus (E)	Gray Mouse Lemur	[76]
M. rufus (E)	Brown Mouse Lemur	[42]
Subfamily Phanerinae		
Genus *Phaner*		
P. furcifer (E)	Fork-marked Lemur	[42]
Family Lemuridae	Large Lemurs	[16]
Genus *Eulemur*	True Lemurs	[76]
E. coronatus (E)	Crowned Lemur	[42]
E. fulvus (E)	Brown Lemur	[42]
E. macaco (E)	Black Lemur	[42]
E. mongoz (E)	Mongoose Lemur	[42]
E. rubriventer (E)	Red-bellied Lemur	[42]
Genus *Hapalemur*	Bamboo Lemurs	[76]
H. aureus (E)	Golden Bamboo Lemur	[42]
H. griseus (E)	Bamboo Lemur	[86]
H. simus (E)	Greater Bamboo Lemur	[76]
Genus *Lemur*		
L. catta (E)	Ring-tailed Lemur	[42]
Genus *Varecia*		
V. variegata (E)	Ruffed Lemur	[42]
Family Megaladapidae	Sportive Lemurs	[86]
Genus *Lepilemur*		
L. dorsalis (E)	Gray-backed Sportive Lemur	[76]
L. edwardsi (E)	Milne-Edwards's Sportive Lemur	[42]
L. leucopus (E)	White-footed Sportive Lemur	[42]
L. microdon (E)	Small-toothed Sportive Lemur	[42]
L. mustelinus (E)	Weasel Sportive Lemur	[42]
L. ruficaudatus (E)	Red-tailed Sportive Lemur	[42]
L. septentrionalis (E)	Northern Sportive Lemur	[42]
Family Indridae	Leaping Lemurs	[16]
Genus *Avahi*		
A. laniger (E)	Woolly Lemur	[42]
Genus *Indri*		
I. indri (E)	Indri	[42]

Genus *Propithecus*	Sifakas	[16]
P. diadema (E)	Diademed Sifaka	[42]
P. tattersalli (E)	Golden-crowned Sifaka	[42]
P. verreauxi (E)	Verreaux's Sifaka	[42]
Family Daubentoniidae		
Genus *Daubentonia*		
D. madagascariensis (E)	Aye-aye	[42]
Family Loridae	Angwantibos, Lorises, and Pottos	
Genus *Arctocebus*	Angwantibos	[63]
A. aureus (E)	Golden Angwantibo	[63]
A. calabarensis (E)	Calabar Angwantibo	[63]
Genus *Loris*		
L. tardigradus (Or)	Slender Loris	[16]
Genus *Nycticebus*	Slow Lorises	[16]
N. coucang (P, Or)	Slow Loris	[16]
N. pygmaeus (P, Or)	Pygmy Slow Loris	[16]
Genus *Perodicticus*		
P. potto (E)	Potto	[66]
Family Galagonidae	Galagos	[63]
Genus *Euoticus*	Needle-clawed Galagos	[63]
E. elegantulus (E)	Western Needle-clawed Galago	[63]
E. pallidus (E)	Pallid Needle-clawed Galago	[63]
Genus *Galago*	Lesser Galagos	[63]
G. alleni (E)	Allen's Squirrel Galago	[63]
G. gallarum (E)	Somali Galago	[63]
G. matschiei (E)	Spectacled Galago	[63]
G. moholi (E)	South African Galago	[63]
G. senegalensis (E)	Senegal Galago	[63]
Genus *Galagoides*	Dwarf Galagos	[63]
G. demidoff (E)	Demidoff's Galago	[63]
G. zanzibaricus (E)	Zanzibar Galago	[63]
Genus *Otolemur*	Greater Galagos	[63]
O. crassicaudatus (E)	Greater Galago	[63]
O. garnettii (E)	Small-eared Galago	[63]
Family Tarsiidae	Tarsiers	[16]
Genus *Tarsius*		
T. bancanus (Or)	Western Tarsier	[16]

T. dianae (A)	Diana Tarsier	
T. pumilus (A)	Pygmy Tarsier	[16]
T. spectrum (A)	Spectral Tarsier	[16]
T. syrichta (Or)	Philippine Tarsier	[48]
Family Callitrichidae	Marmosets and Tamarins	[86]
Genus Callimico		
C. goeldii (Neo)	Goeldi's Monkey	[26]
Genus Callithrix	Marmosets	[26]
C. argentata (Neo)	Silvery Marmoset	[26]
C. aurita (Neo)	White-eared Marmoset	[64]
C. flaviceps (Neo)	Buffy-headed Marmoset	[26]
C. geoffroyi (Neo)	Geoffroy's Marmoset	[64]
C. humeralifer (Neo)	Tassel-eared Marmoset	[26]
C. jacchus (Neo)	White-tufted-ear Marmoset	[26]
C. kuhlii (Neo)	Weid's Black-tufted-ear Marmoset	[26]
C. penicillata (Neo)	Black-pencilled Marmoset	[64]
C. pygmaea (Neo)	Pygmy Marmoset	[26]
Genus Leontopithecus	Lion Tamarins	[86]
L. caissara (Neo)	Black-faced Lion Tamarin	[26]
L. chrysomela (Neo)	Golden-headed Lion Tamarin	[26]
L. chrysopygus (Neo)	Golden-rumped Lion Tamarin	[26]
L. rosalia (Neo)	Golden Lion Tamarin	[26]
Genus Saguinus	Tamarins	[26]
S. bicolor (Neo)	Brazilian Bare-faced Tamarin	[26]
S. fuscicollis (Neo)	Saddlebacked Tamarin	[26]
S. geoffroyi (Neo)	Geoffroy's Tamarin	[26]
S. imperator (Neo)	Emperor Tamarin	[26]
S. inustus (Neo)	Mottle-face Tamarin	[26]
S. labiatus (Neo)	Red-chested Mustached Tamarin	[26]
S. leucopus (Neo)	Silvery-brown Bare-face Tamarin	[26]
S. midas (Neo)	Midas Tamarin	[26]
S. mystax (Neo)	Black-chested Mustached Tamarin	[26]

S. *nigricollis* (Neo)	Black-mantled Tamarin	[26]
S. *oedipus* (Neo)	Cotton-top Tamarin	[26]
S. *tripartitus* (Neo)	Golden-mantled Tamarin	[26]
Family Cebidae	New World Monkeys	[86]
Subfamily Alouattinae		
Genus *Alouatta*	Howler Monkeys	[86]
A. *belzebul* (Neo)	Red-handed Howler Monkey	[26]
A. *caraya* (Neo)	Black Howler Monkey	[26]
A. *coibensis* (Neo)	Coiba Howler Monkey	
A. *fusca* (Neo)	Brown Howler Monkey	[26]
A. *palliata* (Nea, Neo)	Mantled Howler Monkey	[26]
A. *pigra* (Nea, Neo)	Mexican Black Howler Monkey	[26]
A. *sara* (Neo)	Bolivian Red Howler Monkey	
A. *seniculus* (Neo)	Red Howler Monkey	[26]
Subfamily Aotinae		
Genus *Aotus*	Night Monkeys	[86]
A. *azarai* (Neo)	Azara's Night Monkey	[105]
A. *brumbacki* (Neo)	Brumback's Night Monkey	[105]
A. *hershkovitzi* (Neo)	Hershkovitz's Night Monkey	[105]
A. *infulatus* (Neo)	Feline Night Monkey	[105]
A. *lemurinus* (Neo)	Lemurine Night Monkey	[105]
A. *miconax* (Neo)	Andean Night Monkey	[105]
A. *nancymaae* (Neo)	Ma's Night Monkey	[105]
A. *nigriceps* (Neo)	Black-headed Night Monkey	[105]
A. *trivirgatus* (Neo)	Northern Night Monkey	[16]
A. *vociferans* (Neo)	Noisy Night Monkey	[26]
Subfamily Atelinae		
Genus *Ateles*	Spider Monkeys	[86]
A. *belzebuth* (Neo)	White-bellied Spider Monkey	[26]
A. *chamek* (Neo)	Chamek Spider Monkey	
A. *fusciceps* (Neo)	Brown-headed Spider Monkey	[98]
A. *geoffroyi* (Nea, Neo)	Central American Spider Monkey	[26]

A. marginatus (Neo)	White-whiskered Spider Monkey	[105]
A. paniscus (Neo)	Black Spider Monkey	[26]
Genus *Brachyteles*		
B. arachnoides (Neo)	Muriqui	[76]
Genus *Lagothrix*	Woolly Monkeys	[86]
L. flavicauda (Neo)	Yellow-tailed Woolly Monkey	[26]
L. lagotricha (Neo)	Humboldt's Woolly Monkey	
Subfamily Callicebinae		
Genus *Callicebus*	Titis	[16]
C. brunneus (Neo)	Brown Titi	
C. caligatus (Neo)	Booted Titi	
C. cinerascens (Neo)	Ashy Titi	
C. cupreus (Neo)	Coppery Titi	
C. donacophilus (Neo)	Bolivian Titi	
C. dubius (Neo)	Dubius Titi	
C. hoffmannsi (Neo)	Hoffmanns's Titi	
C. modestus (Neo)	Modest Titi	
C. moloch (Neo)	Dusky Titi	[26]
C. oenanthe (Neo)	Andean Titi	
C. olallae (Neo)	Olalla's Titi	
C. personatus (Neo)	Masked Titi	[26]
C. torquatus (Neo)	Yellow-handed Titi	[26]
Subfamily Cebinae		
Genus *Cebus*	Capuchins	[86]
C. albifrons (Neo)	White-fronted Capuchin	[26]
C. apella (Neo)	Brown Capuchin	[26]
C. capucinus (Neo)	White-faced Capuchin	[98]
C. olivaceus (Neo)	Weeping Capuchin	[26]
Genus *Saimiri*	Squirrel Monkeys	[16]
S. boliviensis (Neo)	Bolivian Squirrel Monkey	
S. oerstedii (Neo)	Central American Squirrel Monkey	[26]
S. sciureus (Neo)	South American Squirrel Monkey	
S. ustus (Neo)	Bare-eared Squirrel Monkey	[26]
S. vanzolinii (Neo)	Black Squirrel Monkey	

Subfamily Pitheciinae
  Genus *Cacajao*   Uakaris   [86]
    C. calvus (Neo)   Red Uakari   [26]
    C. melanocephalus (Neo)   Black Uakari   [26]
  Genus *Chiropotes*   Bearded Sakis   [86]
    C. albinasus (Neo)   White-nosed Bearded Saki   [26]
    C. satanas (Neo)   Brown-bearded Saki   [26]
  Genus *Pithecia*   Sakis   [86]
    P. aequatorialis (Neo)   Equatorial Saki   [26]
    P. albicans (Neo)   Buffy Saki   [26]
    P. irrorata (Neo)   Gray Monk Saki   [26]
    P. monachus (Neo)   Monk Saki   [26]
    P. pithecia (Neo)   Guianan Saki   [26]
Family Cercopithecidae   Old World Monkeys   [86]
  Subfamily Cercopithecinae
    Genus *Allenopithecus*
      A. nigroviridis (E)   Allen's Swamp Monkey   [66]
    Genus *Cercocebus*   Mangabeys   [86]
      C. agilis (E)   Agile Mangabey   [109]
      C. galeritus (E)   Tana River Mangabey   [16]
      C. torquatus (E)   Red-capped Mangabey   [66]
    Genus *Cercopithecus*   Guenons   [16]
      C. ascanius (E)   Black-cheeked White-nosed
                                        Monkey   [86]
      C. campbelli (E)   Campbell's Monkey   [16]
      C. cephus (E)   Moustached Monkey   [16]
      C. diana (E)   Diana Monkey   [66]
      C. dryas (E)   Dryas Monkey   [16]
      C. erythrogaster (E)   Red-bellied Monkey   [16]
      C. erythrotis (E)   Red-eared Monkey   [16]
      C. hamlyni (E)   Owl-faced Monkey   [16]
      C. lhoesti (E)   L'Hoest's Monkey   [16]
      C. mitis (E)   Blue Monkey   [66]
      C. mona (E)   Mona Monkey   [66]
      C. neglectus (E)   De Brazza's Monkey   [66]
      C. nictitans (E)   White-nosed Guenon   [86]
      C. petaurista (E)   Lesser White-nosed
                                        Monkey   [86]
      C. pogonias (E)   Crowned Guenon   [16]
      C. preussi (E)   Preuss's Monkey   [16]

C. sclateri (E)	Sclater's Guenon	
C. solatus (E)	Sun-tailed Monkey	[16]
C. wolfi (E)	Wolf's Monkey	[16]
Genus *Chlorocebus*		
C. aethiops (E)	Vervet Monkey	[66]
Genus *Erythrocebus*		
E. patas (E)	Patas Monkey	[66]
Genus *Lophocebus*		
L. albigena (E)	Gray-cheeked Mangabey	[66]
Genus *Macaca*	Macaques	[86]
M. arctoides (P, Or)	Stump-tailed Macaque	[16]
M. assamensis (P, Or)	Assam Macaque	[16]
M. cyclopis (P)	Taiwan Macaque	[16]
M. fascicularis (Or, A)	Long-tailed Macaque	[48]
M. fuscata (P)	Japanese Macaque	[16]
M. maura (A)	Moor Macaque	[16]
M. mulatta (P, Or)	Rhesus Monkey	[86]
M. nemestrina (P, Or)	Pigtail Macaque	[16]
M. nigra (A)	Celebes Crested Macaque	
M. ochreata (A)	Booted Macaque	[16]
M. radiata (Or)	Bonnet Macaque	[16]
M. silenus (Or)	Liontail Macaque	[16]
M. sinica (Or)	Toque Macaque	[16]
M. sylvanus (P)	Barbary Macaque	[66]
M. thibetana (P)	Père David's Macaque	[16]
M. tonkeana (A)	Tonkean Macaque	[16]
Genus *Mandrillus*	Forest Baboons	[16]
M. leucophaeus (E)	Drill	[66]
M. sphinx (E)	Mandrill	[66]
Genus *Miopithecus*		
M. talapoin (E)	Talapoin	[66]
Genus *Papio*		
P. hamadryas (P, E)	Hamadryas Baboon	[66]
Genus *Theropithecus*		
T. gelada (E)	Gelada Baboon	[66]
Subfamily Colobinae		
Genus *Colobus*	Black and White Colobus Monkeys	[86]
C. angolensis (E)	Angolan Colobus	[16]
C. guereza (E)	Guereza	[86]

C. polykomos (E)	King Colobus	
C. satanas (E)	Black Colobus	[66]
Genus *Nasalis*	Long-nosed Monkeys	[109]
N. concolor (Or)	Pig-tailed Langur	[17]
N. larvatus (Or)	Proboscis Monkey	[16]
Genus *Presbytis*	Leaf Monkeys	[86]
P. comata (Or)	Grizzled Leaf Monkey	[17]
P. femoralis (Or)	Banded Leaf Monkey	[17]
P. frontata (Or)	White-fronted Leaf Monkey	[17]
P. hosei (Or)	Gray Leaf Monkey	[17]
P. melalophos (Or)	Mitred Leaf Monkey	[17]
P. potenziani (Or)	Mentawai Leaf Monkey	[17]
P. rubicunda (Or)	Red Leaf Monkey	[17]
P. thomasi (Or)	North Sumatran Leaf Monkey	[17]
Genus *Procolobus*	Red Colobus Monkeys	
P. badius (E)	Western Red Colobus	[63]
P. pennantii (E)	Pennant's Red Colobus	[66]
P. preussi (E)	Preuss's Red Colobus	[66]
P. rufomitratus (E)	Tana River Red Colobus	[63]
P. verus (E)	Olive Colobus	[66]
Genus *Pygathrix*	Snub-nosed Monkeys	
P. avunculus (Or)	Tonkin Snub-nosed Monkey	[17]
P. bieti (P)	Black Snub-nosed Monkey	[17]
P. brelichi (P)	Gray Snub-nosed Monkey	[17]
P. nemaeus (Or)	Douc Langur	[17]
P. roxellana (P)	Golden Snub-nosed Monkey	[17]
Genus *Semnopithecus*		
S. entellus (P, Or)	Hanuman Langur	[17]
Genus *Trachypithecus*	Brow-ridged Langurs	[86]
T. auratus (Or)	Javan Langur	[17]
T. cristatus (Or)	Silvered Leaf Monkey	[17]
T. francoisi (P, Or)	François's Leaf Monkey	[17]
T. geei (Or)	Golden Leaf Monkey	[17]
T. johnii (Or)	Hooded Leaf Monkey	[17]
T. obscurus (Or)	Dusky Leaf Monkey	[17]
T. phayrei (P, Or)	Phayre's Leaf Monkey	[17]
T. pileatus (P, Or)	Capped Leaf Monkey	[17]
T. vetulus (Or)	Purple-faced Leaf Monkey	[17]

Family Hylobatidae	Gibbons	[86]
Genus *Hylobates*		
H. agilis (Or)	Agile Gibbon	[20]
H. concolor (P, Or)	Crested Gibbon	[86]
H. gabriellae (Or)	Buff-cheeked Gibbon	[3]
H. hoolock (P, Or)	Hoolock Gibbon	[16]
H. klossii (Or)	Kloss's Gibbon	[16]
H. lar (Or)	White-handed Gibbon	[17]
H. leucogenys (P, Or)	White-cheeked Gibbon	
H. moloch (Or)	Silvery Gibbon	[86]
H. muelleri (Or)	Borneo Gibbon	[20]
H. pileatus (Or)	Pileated Gibbon	[16]
H. syndactylus (Or)	Siamang	[16]
Family Hominidae	Humans and Great Apes	
Genus *Gorilla*		
G. gorilla (E)	Gorilla	[66]
Genus *Homo*		
H. sapiens (All)	Human	[86]
Genus *Pan*	Chimpanzees	[16]
P. paniscus (E)	Pygmy Chimpanzee	[66]
P. troglodytes (E)	Chimpanzee	[86]
Genus *Pongo*		
P. pygmaeus (Or)	Orangutan	[16]
ORDER CARNIVORA	Carnivores	[16]
Family Canidae	Dogs	[30]
Genus *Alopex*		
A. lagopus (Nea, P)	Arctic Fox	[114]
Genus *Atelocynus*		
A. microtis (Neo)	Short-eared Dog	[26]
Genus *Canis*	Dogs, Wolves, Coyotes, and Jackals	[86]
C. adustus (E)	Side-striped Jackal	[30]
C. aureus (P, E, Or)	Golden Jackal	[30]
C. latrans (Nea, Neo)	Coyote	[114]
C. lupus (Nea, P, Or)	Gray Wolf	[114]
C. mesomelas (E)	Black-backed Jackal	[30]
C. rufus (Nea)	Red Wolf	[114]
C. simensis (E)	Simian Jackal	[30]

Genus *Cerdocyon*
    C. thous (Neo)   Crab-eating Fox   [26]
Genus *Chrysocyon*
    C. brachyurus (Neo)   Maned Wolf   [30]
Genus *Cuon*
    C. alpinus (P, Or)   Dhole   [30]
Genus *Dusicyon*
    D. australis (Neo)   Falkland Island Wolf   [16]
Genus *Lycaon*
    L. pictus (E)   African Wild Dog   [30]
Genus *Nyctereutes*
    N. procyonoides (P, Or)   Raccoon Dog   [30]
Genus *Otocyon*
    O. megalotis (E)   Bat-eared Fox   [30]
Genus *Pseudalopex*   South American Foxes   [86]
    P. culpaeus (Neo)   Culpeo   [97]
    P. griseus (Neo)   Argentine Gray Fox   [16]
    P. gymnocercus (Neo)   Pampas Fox   [7]
    P. sechurae (Neo)   Sechura Fox   [16]
    P. vetulus (Neo)   Hoary Fox   [16]
Genus *Speothos*
    S. venaticus (Neo)   Bush Dog   [30]
Genus *Urocyon*   Gray Foxes   [86]
    U. cinereoargenteus (Nea, Neo)   Gray Fox   [114]
    U. littoralis (Nea)   Island Gray Fox   [114]
Genus *Vulpes*   Foxes   [16]
    V. bengalensis (P, Or)   Bengal Fox   [30]
    V. cana (P)   Blandford's Fox   [30]
    V. chama (E)   Cape Fox   [16]
    V. corsac (P)   Corsac Fox   [30]
    V. ferrilata (P, Or)   Tibetan Fox   [30]
    V. pallida (E)   Pale Fox   [30]
    V. rueppelli (P, E)   Rüppel's Fox   [30]
    V. velox (Nea)   Swift Fox   [54]
    V. vulpes (Nea, P, Or)   Red Fox   [114]
    V. zerda (P)   Fennec   [30]
Family Felidae   Cats   [16]
    Subfamily Acinonychinae
    Genus *Acinonyx*
        A. jubatus (P, E, Or)   Cheetah   [16]

Subfamily Felinae

Genus *Caracal*

C. caracal (P, E, Or)	Caracal	[16]

Genus *Catopuma* — Golden Cats

C. badia (Or)	Bay Cat	[16]
C. temminckii (P, Or)	Asiatic Golden Cat	[16]

Genus *Felis* — Small Cats [16]

F. bieti (P)	Chinese Desert Cat	[16]
F. chaus (P, E, Or)	Jungle Cat	[16]
F. margarita (P, E)	Sand Cat	[16]
F. nigripes (E)	Black-footed Cat	[16]
F. silvestris (P, E, Or)	Wild Cat	[16]

Genus *Herpailurus*

H. yaguarondi (Nea, Neo)	Jaguarundi	[114]

Genus *Leopardus* — Spotted Cats

L. pardalis (Nea, Neo)	Ocelot	[114]
L. tigrinus (Neo)	Little Spotted Cat	[38]
L. wiedii (Nea, Neo)	Margay	[114]

Genus *Leptailurus*

L. serval (P, E)	Serval	[16]

Genus *Lynx* — Lynxes [16]

L. canadensis (Nea)	Canada Lynx	[114]
L. lynx (P)	Eurasian Lynx	[16]
L. pardinus (P)	Spanish Lynx	[86]
L. rufus (Nea)	Bobcat	[114]

Genus *Oncifelis* — South American Cats

O. colocolo (Neo)	Pampas Cat	[16]
O. geoffroyi (Neo)	Geoffroy's Cat	[97]
O. guigna (Neo)	Kodkod	[16]

Genus *Oreailurus*

O. jacobita (Neo)	Andean Cat	[16]

Genus *Otocolobus*

O. manul (P, Or)	Pallas's Cat	[16]

Genus *Prionailurus* — Asian Cats

P. bengalensis (P, Or)	Leopard Cat	[16]
P. planiceps (Or)	Flat-headed Cat	[16]
P. rubiginosus (Or)	Rusty-spotted Cat	[16]
P. viverrinus (P, Or)	Fishing Cat	[16]

Genus *Profelis*

P. aurata (E)	African Golden Cat	[16]

Genus *Puma*
    P. concolor (Nea, Neo)   Puma   [26]
Subfamily Pantherinae
  Genus *Neofelis*
    N. nebulosa (P, Or)   Clouded Leopard   [16]
  Genus *Panthera*   Big Cats   [16]
    P. leo (E, Or)   Lion   [16]
    P. onca (Nea, Neo)   Jaguar   [114]
    P. pardus (P, E, Or)   Leopard   [16]
    P. tigris (P, Or)   Tiger   [16]
  Genus *Pardofelis*
    P. marmorata (P, Or)   Marbled Cat   [16]
  Genus *Uncia*
    U. uncia (P, Or)   Snow Leopard   [16]
Family Herpestidae   Mongooses   [16]
Subfamily Galidiinae
  Genus *Galidia*
    G. elegans (E)   Ring-tailed Mongoose   [16]
  Genus *Galidictis*   Striped Mongooses
    G. fasciata (E)   Broad-striped Mongoose   [16]
    G. grandidieri (E)   Giant-striped Mongoose   [106]
  Genus *Mungotictis*
    M. decemlineata (E)   Narrow-striped Mongoose   [16]
  Genus *Salanoia*
    S. concolor (E)   Brown-tailed Mongoose   [109]
Subfamily Herpestinae
  Genus *Atilax*
    A. paludinosus (P, E)   Marsh Mongoose   [106]
  Genus *Bdeogale*   Black-legged Mongooses   [86]
    B. crassicauda (P, E)   Bushy-tailed Mongoose   [106]
    B. jacksoni (E)   Jackson's Mongoose   [106]
    B. nigripes (E)   Black-footed Mongoose   [106]
  Genus *Crossarchus*   Cusimanses   [86]
    C. alexandri (E)   Alexander's Cusimanse   [106]
    C. ansorgei (E)   Ansorge's Cusimanse   [106]
    C. obscurus (E)   Long-nosed Cusimanse   [109]
  Genus *Cynictis*
    C. penicillata (E)   Yellow Mongoose   [106]
  Genus *Dologale*
    D. dybowskii (E)   Pousargues's Mongoose   [106]

Genus *Galerella*   Slender Mongooses

   G. flavescens (E)   Black Slender Mongoose [106]

   G. pulverulenta (E)   Cape Gray Mongoose [16]

   G. sanguinea (E)   Slender Mongoose [106]

   G. swalius (E)   Namaqua Slender Mongoose

Genus *Helogale*   Dwarf Mongooses [16]

   H. hirtula (E)   Desert Dwarf Mongoose [106]

   H. parvula (E)   Dwarf Mongoose [106]

Genus *Herpestes*   Common Mongooses [16]

   H. brachyurus (Or)   Short-tailed Mongoose [106]

   H. edwardsii (P, Or)   Indian Gray Mongoose [106]

   H. ichneumon (P, E)   Egyptian Mongoose [16]

   H. javanicus (P, Or)   Javan Mongoose [106]

   H. naso (E)   Long-nosed Mongoose [106]

   H. palustris (Or)   Bengal Mongoose [106]

   H. semitorquatus (Or)   Collared Mongoose [106]

   H. smithii (Or)   Ruddy Mongoose [106]

   H. urva (P, Or)   Crab-eating Mongoose [106]

   H. vitticollis (Or)   Striped-necked Mongoose [106]

Genus *Ichneumia*

   I. albicauda (P, E)   White-tailed Mongoose [106]

Genus *Liberiictis*

   L. kuhni (E)   Liberian Mongoose [106]

Genus *Mungos*   Banded Mongooses [86]

   M. gambianus (E)   Gambian Mongoose [86]

   M. mungo (E)   Banded Mongoose [106]

Genus *Paracynictis*

   P. selousi (E)   Selous's Mongoose [106]

Genus *Rhynchogale*

   R. melleri (E)   Meller's Mongoose [106]

Genus *Suricata*

   S. suricatta (E)   Meerkat [16]

Family Hyaenidae   Hyenas [16]

  Subfamily Hyaeninae

   Genus *Crocuta*

    C. crocuta (E)   Spotted Hyena [16]

   Genus *Hyaena*

    H. hyaena (P, E, Or)   Striped Hyena [16]

Genus *Parahyaena*		
P. brunnea (E)	Brown Hyena	[16]
Subfamily Protelinae		
Genus *Proteles*		
P. cristatus (P, E)	Aardwolf	[16]
Family Mustelidae	Weasels, Badgers, Skunks, and Otters	[86]
Subfamily Lutrinae		
Genus *Amblonyx*		
A. cinereus (P, Or)	Oriental Small-clawed Otter	[48]
Genus *Aonyx*	Clawless Otters	[16]
A. capensis (E)	African Clawless Otter	[16]
A. congicus (E)	Congo Clawless Otter	[16]
Genus *Enhydra*		
E. lutris (Oc, Nea, P)	Sea Otter	[114]
Genus *Lontra*	New World River Otters	[86]
L. canadensis (Nea)	Northern River Otter	[114]
L. felina (Neo)	Marine Otter	[109]
L. longicaudis (Nea, Neo)	Neotropical River Otter	[86]
L. provocax (Neo)	Southern River Otter	[109]
Genus *Lutra*	Old World River Otters	[86]
L. lutra (P, Or)	European Otter	[16]
L. maculicollis (E)	Spotted-necked Otter	[16]
L. sumatrana (Or)	Hairy-nosed Otter	[16]
Genus *Lutrogale*		
L. perspicillata (P, Or)	Smooth-coated Otter	[16]
Genus *Pteronura*		
P. brasiliensis (Neo)	Giant Otter	[16]
Subfamily Melinae		
Genus *Arctonyx*		
A. collaris (P, Or)	Hog Badger	[106]
Genus *Meles*		
M. meles (P)	Eurasian Badger	[16]
Genus *Melogale*	Ferret-badgers	[16]
M. everetti (Or)	Everett's Ferret-badger	[16]
M. moschata (P, Or)	Chinese Ferret-badger	[16]
M. orientalis (Or)	Javan Ferret-badger	[106]
M. personata (Or)	Burmese Ferret-badger	[16]
Genus *Mydaus*	Stink Badgers	[16]
M. javanensis (Or)	Sunda Stink Badger	[16]
M. marchei (Or)	Palawan Stink Badger	[48]

Subfamily Mellivorinae
  Genus *Mellivora*

M. capensis (P, E, Or)	Honey Badger	[106]

Subfamily Mephitinae
  Genus *Conepatus*

	Hog-nosed Skunks	[16]
C. chinga (Neo)	Molina's Hog-nosed Skunk	[106]
C. humboldtii (Neo)	Humboldt's Hog-nosed Skunk	
C. leuconotus (Nea)	Eastern Hog-nosed Skunk	[114]
C. mesoleucus (Nea, Neo)	Western Hog-nosed Skunk	[114]
C. semistriatus (Nea, Neo)	Striped Hog-nosed Skunk	[16]

  Genus *Mephitis*

	Striped Skunks	[109]
M. macroura (Nea, Neo)	Hooded Skunk	[114]
M. mephitis (Nea)	Striped Skunk	[114]

  Genus *Spilogale*

	Spotted Skunks	[86]
S. putorius (Nea, Neo)	Eastern Spotted Skunk	[114]
S. pygmaea (Nea)	Pygmy Spotted Skunk	[106]

Subfamily Mustelinae
  Genus *Eira*

E. barbara (Nea, Neo)	Tayra	[106]

  Genus *Galictis*

	Grisons	[86]
G. cuja (Neo)	Lesser Grison	[7]
G. vittata (Nea, Neo)	Greater Grison	[106]

  Genus *Gulo*

G. gulo (Nea, P)	Wolverine	[114]

  Genus *Ictonyx*

	Striped Polecats	[86]
I. libyca (P, E)	Saharan Striped Polecat	
I. striatus (E)	Striped Polecat	[16]

  Genus *Lyncodon*

L. patagonicus (Neo)	Patagonian Weasel	[106]

  Genus *Martes*

	Martens	[86]
M. americana (Nea)	American Marten	[114]
M. flavigula (P, Or)	Yellow-throated Marten	[106]
M. foina (P)	Beech Marten	[106]
M. gwatkinsii (Or)	Nilgiri Marten	[106]
M. martes (P)	European Pine Marten	[106]
M. melampus (P)	Japanese Marten	[106]
M. pennanti (Nea)	Fisher	[114]
M. zibellina (P)	Sable	[106]

Genus *Mustela*	Weasels	[16]
M. africana (Neo)	Tropical Weasel	[106]
M. altaica (P, Or)	Mountain Weasel	[106]
M. erminea (Nea, P)	Ermine	[114]
M. eversmannii (P)	Steppe Polecat	[86]
M. felipei (Neo)	Colombian Weasel	[86]
M. frenata (Nea, Neo)	Long-tailed Weasel	[114]
M. kathiah (P, Or)	Yellow-bellied Weasel	[106]
M. lutreola (P)	European Mink	[106]
M. lutreolina (Or)	Indonesian Mountain Weasel	[106]
M. nigripes (Nea)	Black-footed Ferret	[114]
M. nivalis (Nea, P, Or)	Least Weasel	[114]
M. nudipes (Or)	Malayan Weasel	[16]
M. putorius (P)	European Polecat	[86]
M. sibirica (P, Or)	Siberian Weasel	[106]
M. strigidorsa (P, Or)	Black-striped Weasel	[106]
M. vison (Nea)	American Mink	[114]
Genus *Poecilogale*		
P. albinucha (E)	African Striped Weasel	[106]
Genus *Vormela*		
V. peregusna (P)	Marbled Polecat	[106]
Subfamily Taxidiinae		
Genus *Taxidea*		
T. taxus (Nea)	American Badger	[114]
Family Odobenidae		
Genus *Odobenus*		
O. rosmarus (Oc, Nea, P)	Walrus	[114]
Family Otariidae	Sea Lions	[16]
Genus *Arctocephalus*	Southern Fur Seals	[16]
A. australis (Neo)	South American Fur Seal	[99]
A. forsteri (Oc, A)	New Zealand Fur Seal	[99]
A. galapagoensis (Neo)	Galápagos Fur Seal	[99]
A. gazella (Oc)	Antarctic Fur Seal	[99]
A. philippii (Neo)	Juan Fernandez Fur Seal	[99]
A. pusillus (Oc, E, A)	South African Fur Seal	[99]
A. townsendi (Nea)	Guadalupe Fur Seal	[114]
A. tropicalis (Oc)	Subantarctic Fur Seal	[99]
Genus *Callorhinus*		
C. ursinus (Oc, Nea, P)	Northern Fur Seal	[114]

Genus *Eumetopias*

    E. jubatus (Oc, Nea, P)   Steller Sea Lion   [114]

Genus *Neophoca*

    N. cinerea (A)   Australian Sea Lion   [110]

Genus *Otaria*

    O. byronia (Neo)   South American Sea Lion   [16]

Genus *Phocarctos*

    P. hookeri (Oc, A)   New Zealand Sea Lion   [16]

Genus *Zalophus*

    Z. californianus (Oc, Nea, Neo, P)   California Sea Lion   [114]

Family Phocidae   Earless Seals   [16]

Genus *Cystophora*

    C. cristata (Oc, Nea)   Hooded Seal   [114]

Genus *Erignathus*

    E. barbatus (Oc, Nea, P)   Bearded Seal   [114]

Genus *Halichoerus*

    H. grypus (Oc, Nea, P)   Gray Seal   [114]

Genus *Hydrurga*

    H. leptonyx (Oc, Neo, E, A)   Leopard Seal   [99]

Genus *Leptonychotes*

    L. weddellii (Oc, Neo, A)   Weddell Seal   [99]

Genus *Lobodon*

    L. carcinophagus (Oc, Neo, A)   Crabeater Seal   [99]

Genus *Mirounga*   Elephant Seals   [16]

    M. angustirostris (Oc, Nea)   Northern Elephant Seal   [114]

    M. leonina (Oc, Neo)   Southern Elephant Seal   [99]

Genus *Monachus*   Monk Seals   [16]

    M. monachus (Oc, P)   Mediterranean Monk Seal   [99]

    M. schauinslandi (Oc)   Hawaiian Monk Seal   [99]

    M. tropicalis (Nea, Neo)   West Indian Monk Seal   [54]

Genus *Ommatophoca*

    O. rossii (Oc)   Ross Seal   [99]

Genus *Phoca*   Common Seals

    P. caspica (P)   Caspian Seal   [99]

    P. fasciata (Oc, Nea, P)   Ribbon Seal   [114]

    P. groenlandica (Oc, Nea, P)   Harp Seal   [114]

    P. hispida (Oc, Nea, P)   Ringed Seal   [114]

    P. largha (Oc, Nea, P)   Spotted Seal   [114]

    P. sibirica (P)   Baikal Seal   [99]

    P. vitulina (Oc, Nea, P)   Harbor Seal   [114]

Family Procyonidae	Raccoons and relatives	[16]
Subfamily Potosinae		
Genus *Bassaricyon*	Olingos	[16]
B. alleni (Neo)	Allen's Olingo	[109]
B. beddardi (Neo)	Beddard's Olingo	[109]
B. gabbii (Neo)	Olingo	[26]
B. lasius (Neo)	Harris's Olingo	[16]
B. pauli (Neo)	Chiriqui Olingo	[16]
Genus *Potos*		
P. flavus (Nea, Neo)	Kinkajou	[26]
Subfamily Procyoninae		
Genus *Bassariscus*	Ringtails	[86]
B. astutus (Nea)	Ringtail	[114]
B. sumichrasti (Nea, Neo)	Cacomistle	[26]
Genus *Nasua*	Coatis	[26]
N. narica (Nea, Neo)	White-nosed Coati	[114]
N. nasua (Neo)	South American Coati	[26]
Genus *Nasuella*		
N. olivacea (Neo)	Mountain Coati	[109]
Genus *Procyon*	Raccoons	[26]
P. cancrivorus (Neo)	Crab-eating Raccoon	[26]
P. gloveralleni (Neo)	Barbados Raccoon	[109]
P. insularis (Nea)	Tres Marias Raccoon	[109]
P. lotor (Nea, Neo)	Northern Raccoon	[26]
P. maynardi (Nea)	Bahaman Raccoon	[39]
P. minor (Neo)	Guadeloupe Raccoon	[109]
P. pygmaeus (Nea)	Cozumel Raccoon	[109]
Family Ursidae	Bears	[16]
Subfamily Ailurinae		
Genus *Ailuropoda*		
A. melanoleuca (P)	Giant Panda	[16]
Genus *Ailurus*		
A. fulgens (P, Or)	Red Panda	[16]
Subfamily Ursinae		
Genus *Helarctos*		
H. malayanus (P, Or)	Sun Bear	[16]
Genus *Melursus*		
M. ursinus (Or)	Sloth Bear	[16]
Genus *Tremarctos*		
T. ornatus (Neo)	Spectacled Bear	[16]

Genus *Ursus*	Black, Brown, and Polar Bears	[86]
U. americanus (Nea)	American Black Bear	[114]
U. arctos (Nea, P, Or)	Brown Bear	[114]
U. maritimus (Oc, Nea, P)	Polar Bear	[114]
U. thibetanus (P, Or)	Asiatic Black Bear	[16]
Family Viverridae	Civets and relatives	[86]
Subfamily Cryptoproctinae		
Genus *Cryptoprocta*		
C. ferox (E)	Fossa	[106]
Subfamily Euplerinae		
Genus *Eupleres*		
E. goudotii (E)	Falanouc	[106]
Genus *Fossa*		
F. fossana (E)	Malagasy Civet	[106]
Subfamily Hemigalinae		
Genus *Chrotogale*		
C. owstoni (P, Or)	Owston's Palm Civet	[106]
Genus *Cynogale*		
C. bennettii (Or)	Otter Civet	[106]
Genus *Diplogale*		
D. hosei (Or)	Hose's Palm Civet	[106]
Genus *Hemigalus*		
H. derbyanus (Or)	Banded Palm Civet	[106]
Subfamily Nandiniinae		
Genus *Nandinia*		
N. binotata (E)	African Palm Civet	[106]
Subfamily Paradoxurinae		
Genus *Arctictis*		
A. binturong (P, Or)	Binturong	[106]
Genus *Arctogalidia*		
A. trivirgata (P, Or)	Small-toothed Palm Civet	[106]
Genus *Macrogalidia*		
M. musschenbroekii (A)	Sulawesi Palm Civet	[106]
Genus *Paguma*		
P. larvata (P, Or)	Masked Palm Civet	[106]
Genus *Paradoxurus*	Palm Civets	[86]
P. hermaphroditus (P, Or)	Asian Palm Civet	
P. jerdoni (Or)	Jerdon's Palm Civet	[16]
P. zeylonensis (Or)	Golden Palm Civet	[106]

Subfamily Viverrinae
Genus *Civettictis*
    C. civetta (E) — African Civet — [106]
Genus *Genetta* — Genets — [16]
    G. abyssinica (P, E) — Abyssinian Genet — [106]
    G. angolensis (E) — Angolan Genet — [106]
    G. genetta (P, E) — Small-spotted Genet — [106]
    G. johnstoni (E) — Johnston's Genet — [106]
    G. maculata (E) — Panther Genet — [106]
    G. servalina (E) — Servaline Genet — [106]
    G. thierryi (E) — Haussa Genet — [106]
    G. tigrina (E) — Large-spotted Genet — [16]
    G. victoriae (E) — Giant Genet — [106]
Genus *Osbornictis*
    O. piscivora (E) — Aquatic Genet — [106]
Genus *Poiana*
    P. richardsonii (E) — African Linsang — [106]
Genus *Prionodon* — Oriental Linsangs — [86]
    P. linsang (Or) — Banded Linsang — [106]
    P. pardicolor (P, Or) — Spotted Linsang — [106]
Genus *Viverra* — Oriental Civets — [86]
    V. civettina (Or) — Malabar Civet — [106]
    V. megaspila (P, Or) — Large-spotted Civet — [106]
    V. tangalunga (P, Or, A) — Malayan Civet — [106]
    V. zibetha (P, Or) — Large Indian Civet — [106]
Genus *Viverricula*
    V. indica (P, Or, A) — Small Indian Civet — [106]

ORDER CETACEA — Whales, Dolphins, and Porpoises — [16]

Family Balaenidae — Right Whales — [16]
Genus *Balaena*
    B. mysticetus (Oc, Nea, P) — Bowhead — [65]
Genus *Eubalaena* — Right Whales
    E. australis (Oc, Neo, E, A) — Southern Right Whale — [65]
    E. glacialis (Oc, Nea, P) — Northern Right Whale — [65]
Family Balaenopteridae — Rorquals — [16]
Genus *Balaenoptera* — Baleen Whales
    B. acutorostrata (All) — Minke Whale — [65]
    B. borealis (All) — Sei Whale — [65]

B. edeni (All)	Bryde's Whale	[65]
B. musculus (All)	Blue Whale	[65]
B. physalus (All)	Fin Whale	[65]
Genus *Megaptera*		
M. novaeangliae (All)	Humpback Whale	[114]
Family Eschrichtiidae		
Genus *Eschrichtius*		
E. robustus (Oc, Nea, P)	Gray Whale	[114]
Family Neobalaenidae		
Genus *Caperea*		
C. marginata (Oc, Neo, E, A)	Pygmy Right Whale	[65]
Family Delphinidae	Marine Dolphins	[16]
Genus *Cephalorhynchus*	Piebald Dolphins	[86]
C. commersonii (Oc, Neo)	Commerson's Dolphin	[65]
C. eutropia (Oc, Neo)	Black Dolphin	[65]
C. heavisidii (Oc, E)	Heaviside's Dolphin	[65]
C. hectori (A)	Hector's Dolphin	[65]
Genus *Delphinus*		
D. delphis (All)	Short-beaked Saddleback Dolphin	[114]
Genus *Feresa*		
F. attenuata (All)	Pygmy Killer Whale	[65]
Genus *Globicephala*	Pilot Whales	[16]
G. macrorhynchus (All)	Short-finned Pilot Whale	[65]
G. melas (All)	Long-finned Pilot Whale	[65]
Genus *Grampus*		
G. griseus (Oc)	Risso's Dolphin	[65]
Genus *Lagenodelphis*		
L. hosei (Oc)	Fraser's Dolphin	[65]
Genus *Lagenorhynchus*	White-sided and White-beaked Dolphins	[86]
L. acutus (Oc, Nea, P)	Atlantic White-sided Dolphin	[65]
L. albirostris (Oc, Nea, P)	White-beaked Dolphin	[65]
L. australis (Oc, Neo)	Peale's Dolphin	[65]
L. cruciger (Oc)	Hourglass Dolphin	[65]
L. obliquidens (Oc, Nea, P)	Pacific White-sided Dolphin	[65]
L. obscurus (Oc, Neo, E, A)	Dusky Dolphin	[65]

Genus *Lissodelphis*	Right Whale Dolphins	[16]
L. borealis (Oc, P, Nea)	Northern Right Whale Dolphin	[65]
L. peronii (Oc, Neo, E, A)	Southern Right Whale Dolphin	[65]
Genus *Orcaella*		
O. brevirostris (Or, A)	Irrawaddy Dolphin	[65]
Genus *Orcinus*		
O. orca (All)	Killer Whale	[65]
Genus *Peponocephala*		
P. electra (Oc)	Melon-headed Whale	[65]
Genus *Pseudorca*		
P. crassidens (Oc)	False Killer Whale	[65]
Genus *Sotalia*		
S. fluviatilis (Oc, Neo)	Gray Dolphin	[26]
Genus *Sousa*	Humpbacked Dolphins	[16]
S. chinensis (E, Or, A)	Indo-pacific Humpbacked Dolphin	[65]
S. teuszii (E)	Atlantic Humpbacked Dolphin	[65]
Genus *Stenella*	Spinner, Spotted, and Striped Dolphins	[86]
S. attenuata (All)	Pantropical Spotted Dolphin	[65]
S. clymene (Oc, Nea, Neo, E)	Clymene Dolphin	[65]
S. coeruleoalba (All)	Striped Dolphin	[65]
S. frontalis (Oc, Nea, Neo, E)	Atlantic Spotted Dolphin	[65]
S. longirostris (All)	Spinner Dolphin	[65]
Genus *Steno*		
S. bredanensis (Oc)	Rough-toothed Dolphin	[65]
Genus *Tursiops*		
T. truncatus (All)	Bottlenosed Dolphin	[65]
Family Monodontidae	White Whales	[16]
Genus *Delphinapterus*		
D. leucas (Oc, Nea, P)	Beluga	[65]
Genus *Monodon*		
M. monoceros (Oc, Nea, P)	Narwhal	[65]
Family Phocoenidae	Porpoises	[16]
Genus *Australophocaena*		
A. dioptrica (Oc, Neo, A)	Spectacled Porpoise	[65]

Genus *Neophocaena*		
N. phocaenoides (P, Or)	Finless Porpoise	[65]
Genus *Phocoena*	Harbor Porpoises	[86]
P. phocoena (Oc, Nea, P)	Harbor Porpoise	[65]
P. sinus (Oc, Nea)	Vaquita	[65]
P. spinipinnis (Oc, Neo)	Burmeister's Porpoise	[65]
Genus *Phocoenoides*		
P. dalli (Oc, Nea, P)	Dall's Porpoise	[65]
Family Physeteridae	Sperm Whales	[16]
Genus *Kogia*	Small Sperm Whales	
K. breviceps (Oc)	Pygmy Sperm Whale	[65]
K. simus (Oc)	Dwarf Sperm Whale	[65]
Genus *Physeter*		
P. catodon (Oc)	Sperm Whale	[65]
Family Platanistidae	River Dolphins	[16]
Genus *Inia*		
I. geoffrensis (Neo)	Pink River Dolphin	[26]
Genus *Lipotes*		
L. vexillifer (P)	Yangtze River Dolphin	[65]
Genus *Platanista*	Ganges and Indus Dolphins	[86]
P. gangetica (Or)	Ganges River Dolphin	[65]
P. minor (P)	Indus River Dolphin	[65]
Genus *Pontoporia*		
P. blainvillei (Neo)	Franciscana	[16]
Family Ziphiidae	Beaked Whales	[16]
Genus *Berardius*	Giant Beaked Whales	
B. arnuxii (Oc)	Arnoux's Beaked Whale	[65]
B. bairdii (Oc)	Baird's Beaked Whale	[65]
Genus *Hyperoodon*	Bottlenose Whales	[86]
H. ampullatus (Oc)	Northern Bottlenose Whale	[65]
H. planifrons (Oc)	Southern Bottlenose Whale	[65]
Genus *Indopacetus*		
I. pacificus (Oc)	Longman's Beaked Whale	[65]
Genus *Mesoplodon*	Beaked Whales	[16]
M. bidens (Oc)	Sowerby's Beaked Whale	[65]
M. bowdoini (Oc)	Andrew's Beaked Whale	[65]
M. carlhubbsi (Oc)	Hubbs's Beaked Whale	[65]
M. densirostris (Oc)	Blainville's Beaked Whale	[65]

M. europaeus (Oc)	Gervais's Beaked Whale	[65]
M. ginkgodens (Oc)	Ginkgo-toothed Beaked Whale	[65]
M. grayi (Oc)	Gray's Beaked Whale	[65]
M. hectori (Oc)	Hector's Beaked Whale	[65]
M. layardii (Oc)	Strap-toothed Whale	[65]
M. mirus (Oc)	True's Beaked Whale	[65]
M. peruvianus (Oc)	Pygmy Beaked Whale	[95]
M. stejnegeri (Oc)	Stejneger's Beaked Whale	[65]
Genus *Tasmacetus*		
T. shepherdi (Oc)	Shepherd's Beaked Whale	[65]
Genus *Ziphius*		
Z. cavirostris (Oc)	Cuvier's Beaked Whale	[65]

ORDER SIRENIA	Manatees, Dugongs, and Sea Cows	[86]
Family Dugongidae	Dugongs and Sea Cows	[86]
Genus *Dugong*		
D. dugon (Oc, P, E, Or, A)	Dugong	[16]
Genus *Hydrodamalis*		
H. gigas (Oc, P)	Steller's Sea Cow	[16]
Family Trichechidae	Manatees	[16]
Genus *Trichechus*		
T. inunguis (Neo)	Amazonian Manatee	[16]
T. manatus (Nea, Neo)	West Indian Manatee	[98]
T. senegalensis (E)	African Manatee	[16]

ORDER PROBOSCIDEA	Elephants	[16]
Family Elephantidae		
Genus *Elephas*		
E. maximus (P, Or)	Asiatic Elephant	[86]
Genus *Loxodonta*		
L. africana (E)	African Elephant	[16]

ORDER PERISSODACTYLA	Odd-toed Ungulates	[16]
Family Equidae	Horses, Zebras, and Asses	[86]
Genus *Equus*		
E. asinus (P, E)	Ass	
E. burchellii (E)	Burchell's Zebra	[16]
E. caballus (P)	Horse	

E. grevyi (E)	Grevy's Zebra	[16]
E. hemionus (P)	Kulan	[109]
E. kiang (P, Or)	Kiang	[16]
E. onager (P, Or)	Onager	[16]
E. quagga (E)	Quagga	[16]
E. zebra (E)	Mountain Zebra	[16]
Family Tapiridae	Tapirs	[16]
Genus *Tapirus*		
T. bairdii (Nea, Neo)	Baird's Tapir	[26]
T. indicus (Or)	Malayan Tapir	[16]
T. pinchaque (Neo)	Mountain Tapir	[16]
T. terrestris (Neo)	South American Tapir	[109]
Family Rhinocerotidae	Rhinoceroses	[16]
Genus *Ceratotherium*		
C. simum (E)	White Rhinoceros	[16]
Genus *Dicerorhinus*		
D. sumatrensis (Or)	Sumatran Rhinoceros	[16]
Genus *Diceros*		
D. bicornis (E)	Black Rhinoceros	[16]
Genus *Rhinoceros*	Asian One-horned Rhinoceroses	[86]
R. sondaicus (Or)	Javan Rhinoceros	[16]
R. unicornis (Or)	Indian Rhinoceros	[16]

ORDER HYRACOIDEA

ORDER HYRACOIDEA	Hyraxes	[63]
Family Procaviidae		
Genus *Dendrohyrax*	Tree Hyraxes	[63]
D. arboreus (E)	Southern Tree Hyrax	[63]
D. dorsalis (E)	Western Tree Hyrax	[63]
D. validus (E)	Eastern Tree Hyrax	[63]
Genus *Heterohyrax*	Bush Hyraxes	[63]
H. antineae (P)	Hoggar Hyrax	[63]
H. brucei (P, E)	Yellow-spotted Hyrax	[63]
Genus *Procavia*		
P. capensis (P, E)	Rock Hyrax	[63]

ORDER TUBULIDENTATA

Family Orycteropodidae		
Genus *Orycteropus*		
O. afer (E)	Aardvark	[16]

ORDER ARTIODACTYLA	Even-toed Ungulates	[16]
Family Suidae	Pigs and Hogs	[86]
Subfamily Babyrousinae		
Genus *Babyrousa*		
B. *babyrussa* (A)	Babirusa	[16]
Subfamily Phacochoerinae		
Genus *Phacochoerus*	Warthogs	[63]
P. *aethiopicus* (E)	Desert Warthog	[63]
P. *africanus* (E)	Warthog	[63]
Subfamily Suinae		
Genus *Hylochoerus*		
H. *meinertzhageni* (E)	Giant Forest Hog	[16]
Genus *Potamochoerus*	African Bushpigs	[109]
P. *larvatus* (E)	Bushpig	[63]
P. *porcus* (E)	Red River Hog	[63]
Genus *Sus*	Eurasian Pigs	[16]
S. *barbatus* (Or)	Bearded Pig	[88]
S. *bucculentus* (Or)	Viet Nam Warty Pig	[86]
S. *cebifrons* (Or)	Visayan Warty Pig	[48]
S. *celebensis* (A)	Celebes Wild Boar	[86]
S. *heureni* (A)	Flores Warty Pig	
S. *philippensis* (Or)	Philippine Warty Pig	[48]
S. *salvanius* (Or)	Pygmy Hog	[88]
S. *scrofa* (P, Or, A)	Wild Boar	[16]
S. *timoriensis* (A)	Timor Wild Boar	
S. *verrucosus* (Or)	Javan Pig	[17]
Family Tayassuidae	Peccaries	[16]
Genus *Catagonus*		
C. *wagneri* (Neo)	Chacoan Peccary	[7]
Genus *Pecari*		
P. *tajacu* (Nea, Neo)	Collared Peccary	[7]
Genus *Tayassu*		
T. *pecari* (Nea, Neo)	White-lipped Peccary	[26]
Family Hippopotamidae	Hippopotamuses	[16]
Genus *Hexaprotodon*	Pygmy Hippopotamuses	[86]
H. *liberiensis* (E)	Pygmy Hippopotamus	[16]
H. *madagascariensis* (E)	Madagascan Pygmy Hippopotamus	[88]
Genus *Hippopotamus*	Hippopotamuses	[86]
H. *amphibius* (E)	Hippopotamus	[86]

H. lemerlei (E)	Madagascan Dwarf Hippopotamus	[88]
Family Camelidae	Camels and relatives	
Genus *Camelus*	Camels	[86]
C. bactrianus (P)	Bactrian Camel	[16]
C. dromedarius (P)	Dromedary	[86]
Genus *Lama*	Llamas and relatives	
L. glama (Neo)	Llama	[86]
L. guanicoe (Neo)	Guanaco	[16]
L. pacos (Neo)	Alpaca	[86]
Genus *Vicugna*		
V. vicugna (Neo)	Vicugna	[16]
Family Tragulidae	Chevrotains	[16]
Genus *Hyemoschus*		
H. aquaticus (E)	Water Chevrotain	[16]
Genus *Moschiola*		
M. meminna (Or)	Indian Spotted Chevrotain	[16]
Genus *Tragulus*	Asiatic Mouse-deer	[86]
T. javanicus (P, Or)	Lesser Mouse-deer	
T. napu (Or)	Greater Mouse-deer	
Family Giraffidae	Giraffe and Okapi	[16]
Genus *Giraffa*		
G. camelopardalis (E)	Giraffe	[16]
Genus *Okapia*		
O. johnstoni (E)	Okapi	[16]
Family Moschidae	Musk Deer	[16]
Genus *Moschus*		
M. berezovskii (P, Or)	Chinese Forest Musk Deer	[17]
M. chrysogaster (P, Or)	Alpine Musk Deer	
M. fuscus (P, Or)	Dusky Musk Deer	
M. moschiferus (P)	Siberian Musk Deer	[16]
Family Cervidae	Deer	[16]
Subfamily Cervinae		
Genus *Axis*	Axis Deer	[86]
A. axis (Or)	Chital	[86]
A. calamianensis (Or)	Calamian Deer	[20]
A. kuhlii (Or)	Bawean Deer	[17]
A. porcinus (P, Or)	Hog Deer	[16]
Genus *Cervus*	Red Deer	[86]
C. albirostris (P)	Thorold's Deer	[25]

C. alfredi (Or)	Visayan Spotted Deer	[48]
C. duvaucelii (Or)	Barasingha	[86]
C. elaphus (Nea, P, Or)	Elk	[114]
C. eldii (P, Or)	Eld's Deer	[109]
C. mariannus (Or)	Philippine Brown Deer	[48]
C. nippon (P, Or)	Sika Deer	[86]
C. schomburgki (Or)	Schomburgk's Deer	[16]
C. timorensis (Or)	Timor Deer	[17]
C. unicolor (P, Or)	Sambar	[16]
Genus *Dama*	Fallow Deer	[86]
D. dama (P)	Fallow Deer	[16]
D. mesopotamica (P)	Mesopotamian Fallow Deer	[79]
Genus *Elaphurus*		
E. davidianus (P)	Père David's Deer	[16]
Subfamily Hydropotinae		
Genus *Hydropotes*		
H. inermis (P)	Chinese Water Deer	[16]
Subfamily Muntiacinae		
Genus *Elaphodus*		
E. cephalophus (P, Or)	Tufted Deer	[16]
Genus *Muntiacus*	Muntjacs	[16]
M. atherodes (Or)	Bornean Yellow Muntjac	[16]
M. crinifrons (P)	Black Muntjac	[16]
M. feae (P, Or)	Fea's Muntjac	[16]
M. gongshanensis (P, Or)	Gongshan Muntjac	[17]
M. muntjak (P, Or)	Indian Muntjac	[16]
M. reevesi (P)	Reeves's Muntjac	[17]
Subfamily Capreolinae		
Genus *Alces*		
A. alces (Nea, P)	Moose	[114]
Genus *Blastocerus*		
B. dichotomus (Neo)	Marsh Deer	[16]
Genus *Capreolus*	Roe Deer	[16]
C. capreolus (P)	Western Roe Deer	[16]
C. pygargus (P)	Eastern Roe Deer	[16]
Genus *Hippocamelus*	Guemals	[16]
H. antisensis (Neo)	Peruvian Guemal	[16]
H. bisulcus (Neo)	Chilean Guemal	[16]
Genus *Mazama*	Brockets	[16]
M. americana (Nea, Neo)	Red Brocket	[16]

M. bricenii (Neo)	Merioa Brocket	[89]
M. chunyi (Neo)	Dwarf Brocket	[16]
M. gouazoupira (Neo)	Gray Brocket	[97]
M. nana (Neo)	Pygmy Brocket	
M. rufina (Neo)	Little Red Brocket	[16]
Genus *Odocoileus*	White-tailed Deer and	
	Mule Deer	[86]
O. hemionus (Nea)	Mule Deer	[114]
O. virginianus (Nea, Neo)	White-tailed Deer	[114]
Genus *Ozotoceros*		
O. bezoarticus (Neo)	Pampas Deer	[16]
Genus *Pudu*	Pudus	[86]
P. mephistophiles (Neo)	Northern Pudu	[16]
P. puda (Neo)	Southern Pudu	[16]
Genus *Rangifer*		
R. tarandus (Nea, P)	Caribou	[114]
Family Antilocapridae		
Genus *Antilocapra*		
A. americana (Nea)	Pronghorn	[114]
Family Bovidae	Cattle, Antelopes, Sheep,	
	and Goats	[16]
Subfamily Aepycerotinae		
Genus *Aepyceros*		
A. melampus (E)	Impala	[23]
Subfamily Alcetaphinae		
Genus *Alcelaphus*		
A. buselaphus (E)	Hartebeest	[23]
Genus *Connochaetes*	Wildebeests	[16]
C. gnou (E)	Black Wildebeest	[23]
C. taurinus (E)	Blue Wildebeest	[23]
Genus *Damaliscus*	Sassabies	[86]
D. hunteri (E)	Hirola	[63]
D. lunatus (E)	Topi	[63]
D. pygargus (E)	Bontebok	[23]
Genus *Sigmoceros*		
S. lichtensteinii (E)	Lichtenstein's Hartebeest	[16]
Subfamily Antilopinae		
Genus *Ammodorcas*		
A. clarkei (E)	Dibatag	[63]
Genus *Antidorcas*		
A. marsupialis (E)	Springbok	[23]

Genus *Antilope*		
A. *cervicapra* (P, Or)	Blackbuck	[16]
Genus *Dorcatragus*		
D. *megalotis* (E)	Beira	[109]
Genus *Gazella*	Gazelles	[16]
G. *arabica* (P)	Arabian Gazelle	[79]
G. *bennettii* (P, Or)	Indian Gazelle	[16]
G. *bilkis* (P)	Queen of Sheba's Gazelle	[62]
G. *cuvieri* (P)	Cuvier's Gazelle	[63]
G. *dama* (E)	Dama Gazelle	[23]
G. *dorcas* (P, E)	Dorcas Gazelle	[16]
G. *gazella* (P)	Mountain Gazelle	[16]
G. *granti* (E)	Grant's Gazelle	[16]
G. *leptoceros* (P, E)	Rhim Gazelle	[63]
G. *rufifrons* (E)	Red-fronted Gazelle	[16]
G. *rufina* (P)	Red Gazelle	[16]
G. *saudiya* (P)	Saudi Gazelle	[62]
G. *soemmerringii* (E)	Sömmerring's Gazelle	[63]
G. *spekei* (E)	Speke's Gazelle	[16]
G. *subgutturosa* (P)	Goitered Gazelle	[16]
G. *thomsonii* (E)	Thomson's Gazelle	[16]
Genus *Litocranius*		
L. *walleri* (E)	Gerenuk	[16]
Genus *Madoqua*	Dik-diks	[16]
M. *guentheri* (E)	Günther's Dik-dik	[16]
M. *kirkii* (E)	Kirk's Dik-dik	[23]
M. *piacentinii* (E)	Silver Dik-dik	[63]
M. *saltiana* (E)	Salt's Dik-dik	[16]
Genus *Neotragus*	Dwarf Antelopes	[86]
N. *batesi* (E)	Dwarf Antelope	[63]
N. *moschatus* (E)	Suni	[23]
N. *pygmaeus* (E)	Royal Antelope	[86]
Genus *Oreotragus*		
O. *oreotragus* (E)	Klipspringer	[23]
Genus *Ourebia*		
O. *ourebi* (E)	Oribi	[23]
Genus *Pantholops*		
P. *hodgsonii* (P, Or)	Chiru	[16]
Genus *Procapra*	Central Asian Gazelles	[86]
P. *gutturosa* (P)	Mongolian Gazelle	[16]

P. picticaudata (P, Or)	Tibetan Gazelle	[16]
P. przewalskii (P)	Przewalski's Gazelle	[16]
Genus *Raphicerus*	Steenboks and Grysboks	[86]
R. campestris (E)	Steenbok	[23]
R. melanotis (E)	Cape Grysbok	[63]
R. sharpei (E)	Sharpe's Grysbok	[23]
Genus *Saiga*		
S. tatarica (P)	Saiga	[16]
Subfamily Bovinae		
Genus *Bison*	Bison	[86]
B. bison (Nea)	American Bison	[114]
B. bonasus (P)	European Bison	[16]
Genus *Bos*	Oxen	[16]
B. frontalis (P, Or)	Gaur	[16]
B. grunniens (P, Or)	Yak	[16]
B. javanicus (Or)	Banteng	[16]
B. sauveli (Or)	Kouprey	[16]
B. taurus (P)	Aurochs	[16]
Genus *Boselaphus*		
B. tragocamelus (P, Or)	Nilgai	[16]
Genus *Bubalus*	Asiatic Buffaloes	[16]
B. bubalis (Or)	Water Buffalo	[16]
B. depressicornis (A)	Anoa	[109]
B. mephistopheles (P)	Short-horned Water Buffalo	
B. mindorensis (Or)	Tamaraw	[48]
B. quarlesi (A)	Mountain Anoa	[16]
Genus *Syncerus*		
S. caffer (E)	African Buffalo	[16]
Genus *Taurotragus*	Elands	[86]
T. derbianus (E)	Giant Eland	[16]
T. oryx (E)	Eland	[63]
Genus *Tetracerus*		
T. quadricornis (Or)	Four-horned Antelope	[16]
Genus *Tragelaphus*	Spiral-horned Bovines	[63]
T. angasii (E)	Nyala	[16]
T. buxtoni (E)	Mountain Nyala	[16]
T. eurycerus (E)	Bongo	[16]
T. imberbis (P, E)	Lesser Kudu	[23]
T. scriptus (E)	Bushbuck	[23]

T. spekii (E)	Sitatunga	[23]
T. strepsiceros (E)	Greater Kudu	[63]
Subfamily Caprinae		
Genus *Ammotragus*		
A. lervia (P, E)	Aoudad	[86]
Genus *Budorcas*		
B. taxicolor (P, Or)	Takin	[16]
Genus *Capra*	Goats	[16]
C. caucasica (P)	West Caucasian Tur	[16]
C. cylindricornis (P)	East Caucasian Tur	[16]
C. falconeri (P, Or)	Markhor	[16]
C. hircus (P)	Domestic Goat	[86]
C. ibex (P)	Ibex	[16]
C. nubiana (P, E)	Nubian Ibex	[109]
C. pyrenaica (P)	Spanish Ibex	[109]
C. sibirica (P, Or)	Siberian Ibex	[109]
C. walie (E)	Walia Ibex	[86]
Genus *Hemitragus*	Tahrs	[16]
H. hylocrius (Or)	Nilgiri Tahr	[16]
H. jayakari (P)	Arabian Tahr	[16]
H. jemlahicus (P, Or)	Himalayan Tahr	[16]
Genus *Naemorhedus*	Gorals	[86]
N. baileyi (P, Or)	Red Goral	[16]
N. caudatus (P, Or)	Chinese Goral	[17]
N. crispus (P)	Japanese Serow	[16]
N. goral (P, Or)	Goral	[16]
N. sumatraensis (P, Or)	Serow	[109]
N. swinhoei (P)	Taiwan Serow	[17]
Genus *Oreamnos*		
O. americanus (Nea)	Mountain Goat	[114]
Genus *Ovibos*		
O. moschatus (Nea)	Muskox	[114]
Genus *Ovis*	Sheep	[16]
O. ammon (P, Or)	Argali	[16]
O. aries (P)	Mouflon	[109]
O. canadensis (Nea)	Bighorn Sheep	[114]
O. dalli (Nea)	Dall's Sheep	[114]
O. nivicola (P)	Snow Sheep	[109]
O. vignei (P, Or)	Urial	[16]

Genus *Pseudois*	Bharals	[86]
P. nayaur (P, Or)	Bharal	[16]
P. schaeferi (P)	Dwarf Bharal	
Genus *Rupicapra*	Chamois	[86]
R. pyrenaica (P)	Pyrenean Chamois	[16]
R. rupicapra (P)	Chamois	[16]
Subfamily Cephalophinae		
Genus *Cephalophus*	Forest Duikers	[16]
C. adersi (E)	Aders's Duiker	[16]
C. callipygus (E)	Peters's Duiker	[16]
C. dorsalis (E)	Bay Duiker	[16]
C. harveyi (E)	Harvey's Duiker	[63]
C. jentinki (E)	Jentink's Duiker	[16]
C. leucogaster (E)	White-bellied Duiker	[63]
C. maxwellii (E)	Maxwell's Duiker	[16]
C. monticola (E)	Blue Duiker	[16]
C. natalensis (E)	Natal Duiker	[63]
C. niger (E)	Black Duiker	[16]
C. nigrifrons (E)	Black-fronted Duiker	[16]
C. ogilbyi (E)	Ogilby's Duiker	[16]
C. rubidus (E)	Ruwenzori Duiker	[63]
C. rufilatus (E)	Red-flanked Duiker	[16]
C. silvicultor (E)	Yellow-backed Duiker	[16]
C. spadix (E)	Abbott's Duiker	[16]
C. weynsi (E)	Weyns's Duiker	[23]
C. zebra (E)	Zebra Duiker	[63]
Genus *Sylvicapra*		
S. grimmia (E)	Bush Duiker	[63]
Subfamily Hippotraginae		
Genus *Addax*		
A. nasomaculatus (E)	Addax	[16]
Genus *Hippotragus*	Roan and Sable Antelopes	[86]
H. equinus (E)	Roan Antelope	[23]
H. leucophaeus (E)	Blue Buck	[86]
H. niger (E)	Sable Antelope	[23]
Genus *Oryx*	Oryxes	[86]
O. dammah (E)	Scimitar-horned Oryx	[22]
O. gazella (E)	Gemsbok	[109]
O. leucoryx (P)	Arabian Oryx	[16]

Subfamily Peleinae
    Genus *Pelea*
        P. capreolus (E)   Common Rhebok   [109]
Subfamily Reduncinae
    Genus *Kobus*   Kobs   [63]
        K. ellipsiprymnus (E)   Waterbuck   [23]
        K. kob (E)   Kob   [16]
        K. leche (E)   Lechwe   [23]
        K. megaceros (E)   Nile Lechwe   [16]
        K. vardonii (E)   Puku   [23]
    Genus *Redunca*   Reedbucks   [16]
        R. arundinum (E)   Southern Reedbuck   [23]
        R. fulvorufula (E)   Mountain Reedbuck   [23]
        R. redunca (E)   Bohar Reedbuck   [16]

ORDER PHOLIDOTA   Pangolins   [16]
Family Manidae
    Genus *Manis*
        M. crassicaudata (P, Or)   Indian Pangolin   [16]
        M. gigantea (E)   Giant Pangolin   [109]
        M. javanica (Or)   Malayan Pangolin   [48]
        M. pentadactyla (P, Or)   Chinese Pangolin   [16]
        M. temminckii (E)   Ground Pangolin   [63]
        M. tetradactyla (E)   Long-tailed Pangolin   [16]
        M. tricuspis (E)   Tree Pangolin   [16]

ORDER RODENTIA   Rodents   [16]
Family Aplodontidae
    Genus *Aplodontia*
        A. rufa (Nea)   Mountain Beaver   [114]
Family Sciuridae   Squirrels   [16]
    Subfamily Sciurinae
    Genus *Ammospermophilus*   Antelope Squirrels   [16]
        A. harrisii (Nea)   Harris's Antelope Squirrel   [114]
        A. insularis (Nea)   Espiritu Santo Island
                Antelope Squirrel   [38]
        A. interpres (Nea)   Texas Antelope Squirrel   [114]
        A. leucurus (Nea)   White-tailed Antelope
                Squirrel   [114]
        A. nelsoni (Nea)   Nelson's Antelope Squirrel   [114]

Genus *Atlantoxerus*		
A. *getulus* (P)	Barbary Ground Squirrel	[16]
Genus *Callosciurus*	Oriental Tree Squirrels	[16]
C. *adamsi* (Or)	Ear-spot Squirrel	[17]
C. *albescens* (Or)	Kloss Squirrel	[20]
C. *baluensis* (Or)	Kinabalu Squirrel	[16]
C. *caniceps* (Or)	Gray-bellied Squirrel	[16]
C. *erythraeus* (P, Or)	Pallas's Squirrel	[109]
C. *finlaysonii* (Or)	Finlayson's Squirrel	[16]
C. *inornatus* (P, Or)	Inornate Squirrel	
C. *melanogaster* (Or)	Mentawai Squirrel	[20]
C. *nigrovittatus* (Or)	Black-striped Squirrel	[17]
C. *notatus* (Or, A)	Plantain Squirrel	[16]
C. *orestes* (Or)	Borneo Black-banded Squirrel	[20]
C. *phayrei* (P, Or)	Phayre's Squirrel	
C. *prevostii* (Or)	Prevost's Squirrel	[16]
C. *pygerythrus* (P, Or)	Irrawaddy Squirrel	[16]
C. *quinquestriatus* (P, Or)	Anderson's Squirrel	[16]
Genus *Cynomys*	Prairie Dogs	[16]
C. *gunnisoni* (Nea)	Gunnison's Prairie Dog	[114]
C. *leucurus* (Nea)	White-tailed Prairie Dog	[114]
C. *ludovicianus* (Nea)	Black-tailed Prairie Dog	[114]
C. *mexicanus* (Nea)	Mexican Prairie Dog	[16]
C. *parvidens* (Nea)	Utah Prairie Dog	[114]
Genus *Dremomys*	Red-cheeked Squirrels	[86]
D. *everetti* (Or)	Bornean Mountain Ground Squirrel	[16]
D. *lokriah* (P, Or)	Orange-bellied Himalayan Squirrel	[16]
D. *pernyi* (P, Or)	Perny's Long-nosed Squirrel	[16]
D. *pyrrhomerus* (P, Or)	Red-hipped Squirrel	
D. *rufigenis* (P, Or)	Asian Red-cheeked Squirrel	[109]
Genus *Epixerus*	African Palm Squirrels	[16]
E. *ebii* (E)	Western Palm Squirrel	[63]
E. *wilsoni* (E)	Biafran Palm Squirrel	[63]
Genus *Exilisciurus*	Pygmy Squirrels	[86]
E. *concinnus* (Or)	Philippine Pygmy Squirrel	[45]

E. exilis (Or)	Least Pygmy Squirrel	[17]
E. whiteheadi (Or)	Tufted Pygmy Squirrel	[17]
Genus *Funambulus*	Asiatic Palm Squirrels	
F. layardi (Or)	Layard's Palm Squirrel	
F. palmarum (Or)	Indian Palm Squirrel	[16]
F. pennantii (P, Or)	Northern Palm Squirrel	[16]
F. sublineatus (Or)	Dusky Palm Squirrel	
F. tristriatus (Or)	Jungle Palm Squirrel	
Genus *Funisciurus*	Rope Squirrels	[63]
F. anerythrus (E)	Thomas's Rope Squirrel	[63]
F. bayonii (E)	Lunda Rope Squirrel	[63]
F. carruthersi (E)	Carruther's Mountain Squirrel	[63]
F. congicus (E)	Congo Rope Squirrel	[63]
F. isabella (E)	Lady Burton's Rope Squirrel	[63]
F. lemniscatus (E)	Ribboned Rope Squirrel	[63]
F. leucogenys (E)	Red-cheeked Rope Squirrel	[63]
F. pyrropus (E)	Fire-footed Rope Squirrel	[63]
F. substriatus (E)	Kintampo Rope Squirrel	[63]
Genus *Glyphotes*		
G. simus (Or)	Sculptor Squirrel	[17]
Genus *Heliosciurus*	Sun Squirrels	[16]
H. gambianus (E)	Gambian Sun Squirrel	[16]
H. mutabilis (E)	Mutable Sun Squirrel	[63]
H. punctatus (E)	Small Sun Squirrel	
H. rufobrachium (E)	Red-legged Sun Squirrel	[16]
H. ruwenzorii (E)	Ruwenzori Sun Squirrel	[16]
H. undulatus (E)	Zanj Sun Squirrel	[63]
Genus *Hyosciurus*	Sulawesi Long-nosed Squirrels	[17]
H. heinrichi (A)	Montane Long-nosed Squirrel	[17]
H. ileile (A)	Lowland Long-nosed Squirrel	[17]
Genus *Lariscus*	Striped Ground Squirrels	[17]
L. hosei (Or)	Four-striped Ground Squirrel	[16]
L. insignis (Or)	Three-striped Ground Squirrel	[16]

L. niobe (Or)	Niobe Ground Squirrel	
L. obscurus (Or)	Mentawai Three-striped	
	Squirrel	[17]
Genus *Marmota*	Marmots	[16]
M. baibacina (P)	Gray Marmot	
M. bobak (P)	Bobak Marmot	[16]
M. broweri (Nea)	Alaska Marmot	[114]
M. caligata (Nea)	Hoary Marmot	[114]
M. camtschatica (P)	Black-capped Marmot	[16]
M. caudata (P, Or)	Long-tailed Marmot	[16]
M. flaviventris (Nea)	Yellow-bellied Marmot	[114]
M. himalayana (P, Or)	Himalayan Marmot	[16]
M. marmota (P)	Alpine Marmot	[16]
M. menzbieri (P)	Menzbier's Marmot	[16]
M. monax (Nea)	Woodchuck	[114]
M. olympus (Nea)	Olympic Marmot	[114]
M. sibirica (P)	Tarbagan Marmot	
M. vancouverensis (Nea)	Vancouver Marmot	[114]
Genus *Menetes*		
M. berdmorei (P, Or)	Indochinese Ground	
	Squirrel	[16]
Genus *Microsciurus*	Dwarf Squirrels	[26]
M. alfari (Neo)	Central American Dwarf	
	Squirrel	[26]
M. flaviventer (Neo)	Amazon Dwarf Squirrel	[26]
M. mimulus (Neo)	Western Dwarf Squirrel	[26]
M. santanderensis (Neo)	Santander Dwarf Squirrel	
Genus *Myosciurus*		
M. pumilio (E)	African Pygmy Squirrel	[63]
Genus *Nannosciurus*		
N. melanotis (Or)	Black-eared Squirrel	[86]
Genus *Paraxerus*	Bush Squirrels	[63]
P. alexandri (E)	Alexander's Bush Squirrel	[16]
P. boehmi (E)	Boehm's Bush Squirrel	[16]
P. cepapi (E)	Smith's Bush Squirrel	[16]
P. cooperi (E)	Cooper's Mountain Squirrel	[63]
P. flavovittis (E)	Striped Bush Squirrel	[16]
P. lucifer (E)	Black and Red Bush	
	Squirrel	[16]
P. ochraceus (E)	Ochre Bush Squirrel	[63]

P. palliatus (E)	Red Bush Squirrel	[16]
P. poensis (E)	Green Bush Squirrel	[63]
P. vexillarius (E)	Svynnerton's Bush Squirrel	[16]
P. vincenti (E)	Vincent's Bush Squirrel	[16]
Genus *Prosciurillus*	Sulawesi Dwarf Squirrels	[16]
P. abstrusus (A)	Secretive Dwarf Squirrel	
P. leucomus (A)	Whitish Dwarf Squirrel	
P. murinus (A)	Celebes Dwarf Squirrel	[109]
P. weberi (A)	Weber's Dwarf Squirrel	
Genus *Protoxerus*	African Giant Squirrels	[16]
P. aubinnii (E)	Slender-tailed Squirrel	[63]
P. stangeri (E)	Forest Giant Squirrel	[16]
Genus *Ratufa*	Oriental Giant Squirrels	[16]
R. affinis (Or)	Pale Giant Squirrel	[17]
R. bicolor (P, Or)	Black Giant Squirrel	[16]
R. indica (Or)	Indian Giant Squirrel	[16]
R. macroura (Or)	Sri Lankan Giant Squirrel	[17]
Genus *Rheithrosciurus*		
R. macrotis (Or)	Tufted Ground Squirrel	[16]
Genus *Rhinosciurus*		
R. laticaudatus (Or)	Shrew-faced Squirrel	[17]
Genus *Rubrisciurus*		
R. rubriventer (A)	Sulawesi Giant Squirrel	[17]
Genus *Sciurillus*		
S. pusillus (Neo)	Neotropical Pygmy Squirrel	[109]
Genus *Sciurotamias*	Asian Rock Squirrels	[86]
S. davidianus (P)	Père David's Rock Squirrel	[17]
S. forresti (P)	Forrest's Rock Squirrel	[17]
Genus *Sciurus*	Tree Squirrels	[86]
S. aberti (Nea)	Abert's Squirrel	[114]
S. aestuans (Neo)	Guianan Squirrel	[26]
S. alleni (Nea)	Allen's Squirrel	[38]
S. anomalus (P)	Caucasian Squirrel	[109]
S. arizonensis (Nea)	Arizona Gray Squirrel	[114]
S. aureogaster (Nea, Neo)	Red-bellied Squirrel	[109]
S. carolinensis (Nea)	Eastern Gray Squirrel	[114]
S. colliaei (Nea)	Collie's Squirrel	[38]
S. deppei (Nea, Neo)	Deppe's Squirrel	[26]
S. flammifer (Neo)	Fiery Squirrel	
S. gilvigularis (Neo)	Yellow-throated Squirrel	

S. granatensis (Neo)	Red-tailed Squirrel	[26]
S. griseus (Nea)	Western Gray Squirrel	[114]
S. ignitus (Neo)	Bolivian Squirrel	[26]
S. igniventris (Neo)	Northern Amazon Red Squirrel	[26]
S. lis (P)	Japanese Squirrel	[16]
S. nayaritensis (Nea)	Mexican Fox Squirrel	[114]
S. niger (Nea)	Eastern Fox Squirrel	[114]
S. oculatus (Nea)	Peters's Squirrel	[38]
S. pucheranii (Neo)	Andean Squirrel	
S. pyrrhinus (Neo)	Junín Red Squirrel	[26]
S. richmondi (Neo)	Richmond's Squirrel	[26]
S. sanborni (Neo)	Sanborn's Squirrel	[26]
S. spadiceus (Neo)	Southern Amazon Red Squirrel	[26]
S. stramineus (Neo)	Guayaquil Squirrel	[26]
S. variegatoides (Nea, Neo)	Variegated Squirrel	[26]
S. vulgaris (P)	Eurasian Red Squirrel	[16]
S. yucatanensis (Nea, Neo)	Yucatan Squirrel	[26]
Genus *Spermophilopsis*		
S. leptodactylus (P)	Long-clawed Ground Squirrel	[16]
Genus *Spermophilus*	Ground Squirrels	[16]
S. adocetus (Nea)	Tropical Ground Squirrel	[16]
S. alashanicus (P)	Alashan Ground Squirrel	[16]
S. annulatus (Nea)	Ring-tailed Ground Squirrel	[38]
S. armatus (Nea)	Uinta Ground Squirrel	[114]
S. atricapillus (Nea)	Baja California Rock Squirrel	[38]
S. beecheyi (Nea)	California Ground Squirrel	[114]
S. beldingi (Nea)	Belding's Ground Squirrel	[114]
S. brunneus (Nea)	Idaho Ground Squirrel	[114]
S. canus (Nea)	Merriam's Ground Squirrel	[114]
S. citellus (P)	European Ground Squirrel	[109]
S. columbianus (Nea)	Columbian Ground Squirrel	[114]
S. dauricus (P)	Daurian Ground Squirrel	[16]
S. elegans (Nea)	Wyoming Ground Squirrel	[114]
S. erythrogenys (P)	Red-cheeked Ground Squirrel	[109]

S. franklinii (Nea)	Franklin's Ground Squirrel	[114]
S. fulvus (P)	Yellow Ground Squirrel	[109]
S. lateralis (Nea)	Golden-mantled Ground Squirrel	[114]
S. madrensis (Nea)	Sierra Madre Ground Squirrel	[16]
S. major (P)	Russet Ground Squirrel	[109]
S. mexicanus (Nea)	Mexican Ground Squirrel	[114]
S. mohavensis (Nea)	Mohave Ground Squirrel	[114]
S. mollis (Nea)	Piute Ground Squirrel	[114]
S. musicus (P)	Caucasian Mountain Ground Squirrel	[109]
S. parryii (P, Nea)	Arctic Ground Squirrel	[114]
S. perotensis (Nea)	Perote Ground Squirrel	[38]
S. pygmaeus (P)	Little Ground Squirrel	
S. relictus (P)	Tien Shan Ground Squirrel	
S. richardsonii (Nea)	Richardson's Ground Squirrel	[114]
S. saturatus (Nea)	Cascade Golden-mantled Ground Squirrel	[114]
S. spilosoma (Nea)	Spotted Ground Squirrel	[114]
S. suslicus (P)	Speckled Ground Squirrel	[109]
S. tereticaudus (Nea)	Round-tailed Ground Squirrel	[114]
S. townsendii (Nea)	Townsend's Ground Squirrel	[114]
S. tridecemlineatus (Nea)	Thirteen-lined Ground Squirrel	[114]
S. undulatus (P)	Long-tailed Ground Squirrel	
S. variegatus (Nea)	Rock Squirrel	[114]
S. washingtoni (Nea)	Washington Ground Squirrel	[114]
S. xanthoprymnus (P)	Asia Minor Ground Squirrel	[109]
Genus *Sundasciurus*	Sunda Squirrels	[16]
S. brookei (Or)	Brooke's Squirrel	[16]
S. davensis (Or)	Davao Squirrel	
S. fraterculus (Or)	Fraternal Squirrel	
S. hippurus (Or)	Horse-tailed Squirrel	[16]
S. hoogstraali (Or)	Busuanga Squirrel	[20]

S. jentinki (Or)	Jentink's Squirrel	[16]
S. juvencus (Or)	Northern Palawan Tree Squirrel	[48]
S. lowii (Or)	Low's Squirrel	[16]
S. mindanensis (Or)	Mindanao Squirrel	
S. moellendorffi (Or)	Culion Tree Squirrel	[48]
S. philippinensis (Or)	Philippine Tree Squirrel	[48]
S. rabori (Or)	Palawan Montane Squirrel	[17]
S. samarensis (Or)	Samar Squirrel	
S. steerii (Or)	Southern Palawan Tree Squirrel	[48]
S. tenuis (Or)	Slender Squirrel	[16]
Genus Syntheosciurus		
S. brochus (Neo)	Bangs's Mountain Squirrel	[38]
Genus Tamias	Chipmunks	[16]
T. alpinus (Nea)	Alpine Chipmunk	[114]
T. amoenus (Nea)	Yellow-pine Chipmunk	[114]
T. bulleri (Nea)	Buller's Chipmunk	[38]
T. canipes (Nea)	Gray-footed Chipmunk	[114]
T. cinereicollis (Nea)	Gray-collared Chipmunk	[114]
T. dorsalis (Nea)	Cliff Chipmunk	[114]
T. durangae (Nea)	Durango Chipmunk	
T. merriami (Nea)	Merriam's Chipmunk	[114]
T. minimus (Nea)	Least Chipmunk	[114]
T. obscurus (Nea)	California Chipmunk	[114]
T. ochrogenys (Nea)	Yellow-cheeked Chipmunk	[114]
T. palmeri (Nea)	Palmer's Chipmunk	[114]
T. panamintinus (Nea)	Panamint Chipmunk	[114]
T. quadrimaculatus (Nea)	Long-eared Chipmunk	[114]
T. quadrivittatus (Nea)	Colorado Chipmunk	[114]
T. ruficaudus (Nea)	Red-tailed Chipmunk	[114]
T. rufus (Nea)	Hopi Chipmunk	[114]
T. senex (Nea)	Allen's Chipmunk	[114]
T. sibiricus (P)	Siberian Chipmunk	[16]
T. siskiyou (Nea)	Siskiyou Chipmunk	[114]
T. sonomae (Nea)	Sonoma Chipmunk	[114]
T. speciosus (Nea)	Lodgepole Chipmunk	[114]
T. striatus (Nea)	Eastern Chipmunk	[114]
T. townsendii (Nea)	Townsend's Chipmunk	[114]
T. umbrinus (Nea)	Uinta Chipmunk	[114]

Genus *Tamiasciurus*	Red Squirrels	[86]
T. douglasii (Nea)	Douglas's Squirrel	[114]
T. hudsonicus (Nea)	Red Squirrel	[114]
T. mearnsi (Nea)	Mearns's Squirrel	
Genus *Tamiops*	Asiatic Striped Squirrels	[86]
T. macclellandi (P, Or)	Himalayan Striped Squirrel	[16]
T. maritimus (P, Or)	Maritime Striped Squirrel	
T. rodolphei (Or)	Cambodian Striped Squirrel	[16]
T. swinhoei (P, Or)	Swinhoe's Striped Squirrel	[16]
Genus *Xerus*	African Ground Squirrels	[16]
X. erythropus (P, E)	Striped Ground Squirrel	[63]
X. inauris (E)	South African Ground Squirrel	[63]
X. princeps (E)	Damara Ground Squirrel	[63]
X. rutilus (E)	Unstriped Ground Squirrel	[16]
Subfamily Pteromyinae		
Genus *Aeretes*		
A. melanopterus (P)	North Chinese Flying Squirrel	[16]
Genus *Aeromys*	Large Black Flying Squirrels	[86]
A. tephromelas (Or)	Black Flying Squirrel	[16]
A. thomasi (Or)	Thomas's Flying Squirrel	[17]
Genus *Belomys*		
B. pearsonii (P, Or)	Hairy-footed Flying Squirrel	[16]
Genus *Biswamoyopterus*		
B. biswasi (Or)	Namdapha Flying Squirrel	[16]
Genus *Eupetaurus*		
E. cinereus (P, Or)	Woolly Flying Squirrel	[16]
Genus *Glaucomys*	New World Flying Squirrels	[86]
G. sabrinus (Nea)	Northern Flying Squirrel	[114]
G. volans (Nea, Neo)	Southern Flying Squirrel	[114]
Genus *Hylopetes*	Arrow-tailed Flying Squirrels	[86]
H. alboniger (P, Or)	Particolored Flying Squirrel	[16]
H. baberi (P, Or)	Afghan Flying Squirrel	
H. bartelsi (Or)	Bartel's Flying Squirrel	
H. fimbriatus (Or)	Kashmir Flying Squirrel	[16]
H. lepidus (Or)	Gray-cheeked Flying Squirrel	[17]
H. nigripes (Or)	Palawan Flying Squirrel	[20]

H. phayrei (P, Or)	Indochinese Flying Squirrel	
H. sipora (Or)	Sipora Flying Squirrel	[20]
H. spadiceus (Or)	Red-cheeked Flying Squirrel	[17]
H. winstoni (Or)	Sumatran Flying Squirrel	
Genus *Iomys*	Horsfield's Flying Squirrels	[86]
I. horsfieldi (Or)	Javanese Flying Squirrel	[109]
I. sipora (Or)	Mentawi Flying Squirrel	
Genus *Petaurillus*	Pygmy Flying Squirrels	[86]
P. emiliae (Or)	Lesser Pygmy Flying Squirrel	[91]
P. hosei (Or)	Hose's Pygmy Flying Squirrel	[91]
P. kinlochii (Or)	Selangor Pygmy Flying Squirrel	[16]
Genus *Petaurista*	Giant Flying Squirrels	[16]
P. alborufus (P)	Red and White Giant Flying Squirrel	
P. elegans (P, Or)	Spotted Giant Flying Squirrel	[16]
P. leucogenys (P)	Japanese Giant Flying Squirrel	[16]
P. magnificus (Or)	Hodgson's Giant Flying Squirrel	
P. nobilis (Or)	Bhutan Giant Flying Squirrel	
P. petaurista (P, Or)	Red Giant Flying Squirrel	[16]
P. philippensis (P, Or)	Indian Giant Flying Squirrel	[16]
P. xanthotis (P)	Chinese Giant Flying Squirrel	[16]
Genus *Petinomys*	Dwarf Flying Squirrels	[86]
P. crinitus (Or)	Mindanao Flying Squirrel	[48]
P. fuscocapillus (Or)	Travancore Flying Squirrel	[16]
P. genibarbis (Or)	Whiskered Flying Squirrel	[16]
P. hageni (Or)	Hagen's Flying Squirrel	[20]
P. lugens (Or)	Siberut Flying Squirrel	
P. sagitta (Or)	Arrow Flying Squirrel	
P. setosus (Or)	Temminck's Flying Squirrel	[17]
P. vordermanni (Or)	Vordermann's Flying Squirrel	[16]

Genus *Pteromys*	Eurasian Flying Squirrels	
P. momonga (P)	Japanese Flying Squirrel	
P. volans (P)	Siberian Flying Squirrel	[16]
Genus *Pteromyscus*		
P. pulverulentus (Or)	Smoky Flying Squirrel	[16]
Genus *Trogopterus*		
T. xanthipes (P)	Complex-toothed Flying Squirrel	[16]
Family Castoridae	Beavers	[16]
Genus *Castor*		
C. canadensis (Nea)	American Beaver	[114]
C. fiber (P)	Eurasian Beaver	[16]
Family Geomyidae	Pocket Gophers	[16]
Genus *Geomys*	Eastern Pocket Gophers	[86]
G. arenarius (Nea)	Desert Pocket Gopher	[114]
G. bursarius (Nea)	Plains Pocket Gopher	[114]
G. personatus (Nea)	Texas Pocket Gopher	[114]
G. pinetis (Nea)	Southeastern Pocket Gopher	[114]
G. tropicalis (Nea)	Tropical Pocket Gopher	[38]
Genus *Orthogeomys*	Giant Pocket Gophers	[38]
O. cavator (Neo)	Chiriqui Pocket Gopher	[38]
O. cherriei (Neo)	Cherrie's Pocket Gopher	[38]
O. cuniculus (Nea)	Oaxacan Pocket Gopher	[38]
O. dariensis (Neo)	Darien Pocket Gopher	[38]
O. grandis (Nea, Neo)	Giant Pocket Gopher	[109]
O. heterodus (Neo)	Variable Pocket Gopher	[38]
O. hispidus (Nea, Neo)	Hispid Pocket Gopher	[38]
O. lanius (Nea)	Big Pocket Gopher	[38]
O. matagalpae (Neo)	Nicaraguan Pocket Gopher	[38]
O. thaeleri (Neo)	Thaeler's Pocket Gopher	
O. underwoodi (Neo)	Underwood's Pocket Gopher	[38]
Genus *Pappogeomys*	Mexican Pocket Gophers	
P. alcorni (Nea)	Alcorn's Pocket Gopher	[38]
P. bulleri (Nea)	Buller's Pocket Gopher	[38]
P. castanops (Nea)	Yellow-faced Pocket Gopher	[114]
P. fumosus (Nea)	Smoky Pocket Gopher	[38]
P. gymnurus (Nea)	Llano Pocket Gopher	[38]
P. merriami (Nea)	Merriam's Pocket Gopher	[38]

P. neglectus (Nea)	Querétaro Pocket Gopher	[38]
P. tylorhinus (Nea)	Naked-nosed Pocket Gopher	
P. zinseri (Nea)	Zinser's Pocket Gopher	[38]
Genus *Thomomys*	Western Pocket Gophers	[86]
T. bottae (Nea)	Botta's Pocket Gopher	[114]
T. bulbivorus (Nea)	Camas Pocket Gopher	[114]
T. clusius (Nea)	Wyoming Pocket Gopher	[114]
T. idahoensis (Nea)	Idaho Pocket Gopher	[114]
T. mazama (Nea)	Western Pocket Gopher	[114]
T. monticola (Nea)	Mountain Pocket Gopher	[114]
T. talpoides (Nea)	Northern Pocket Gopher	[114]
T. townsendii (Nea)	Townsend's Pocket Gopher	[114]
T. umbrinus (Nea)	Southern Pocket Gopher	[114]
Genus *Zygogeomys*		
Z. trichopus (Nea)	Michoacan Pocket Gopher	[38]
Family Heteromyidae	Pocket Mice, Kangaroo Rats, and Kangaroo Mice	[86]
Subfamily Dipodomyinae		
Genus *Dipodomys*	Kangaroo Rats	[114]
D. agilis (Nea)	Agile Kangaroo Rat	[114]
D. californicus (Nea)	California Kangaroo Rat	[114]
D. compactus (Nea)	Gulf Coast Kangaroo Rat	[114]
D. deserti (Nea)	Desert Kangaroo Rat	[114]
D. elator (Nea)	Texas Kangaroo Rat	[114]
D. elephantinus (Nea)	Big-eared Kangaroo Rat	[114]
D. gravipes (Nea)	San Quintin Kangaroo Rat	[38]
D. heermanni (Nea)	Heermann's Kangaroo Rat	[114]
D. ingens (Nea)	Giant Kangaroo Rat	[114]
D. insularis (Nea)	San Jose Island Kangaroo Rat	[38]
D. margaritae (Nea)	Margarita Island Kangaroo Rat	[38]
D. merriami (Nea)	Merriam's Kangaroo Rat	[114]
D. microps (Nea)	Chisel-toothed Kangaroo Rat	[114]
D. nelsoni (Nea)	Nelson's Kangaroo Rat	[38]
D. nitratoides (Nea)	Fresno Kangaroo Rat	[114]
D. ordii (Nea)	Ord's Kangaroo Rat	[114]
D. panamintinus (Nea)	Panamint Kangaroo Rat	[114]

D. phillipsii (Nea)	Phillips's Kangaroo Rat	[38]
D. spectabilis (Nea)	Banner-tailed Kangaroo Rat	[114]
D. stephensi (Nea)	Stephens's Kangaroo Rat	[114]
D. venustus (Nea)	Narrow-faced Kangaroo Rat	[114]
Genus *Microdipodops*	Kangaroo Mice	[16]
M. megacephalus (Nea)	Dark Kangaroo Mouse	[114]
M. pallidus (Nea)	Pale Kangaroo Mouse	[114]
Subfamily Heteromyinae		
Genus *Heteromys*	Forest Spiny Pocket Mice	[16]
H. anomalus (Neo)	Trinidad Spiny Pocket Mouse	[16]
H. australis (Neo)	Southern Spiny Pocket Mouse	[38]
H. desmarestianus (Nea, Neo)	Desmarest's Spiny Pocket Mouse	[38]
H. gaumeri (Nea, Neo)	Gaumer's Spiny Pocket Mouse	[38]
H. goldmani (Nea, Neo)	Goldman's Spiny Pocket Mouse	[38]
H. nelsoni (Nea)	Nelson's Spiny Pocket Mouse	[38]
H. oresterus (Neo)	Mountain Spiny Pocket Mouse	[38]
Genus *Liomys*	Spiny Pocket Mice	[16]
L. adspersus (Neo)	Panamanian Spiny Pocket Mouse	[38]
L. irroratus (Nea)	Mexican Spiny Pocket Mouse	[114]
L. pictus (Nea, Neo)	Painted Spiny Pocket Mouse	[38]
L. salvini (Nea, Neo)	Salvin's Spiny Pocket Mouse	[38]
L. spectabilis (Nea)	Jaliscan Spiny Pocket Mouse	[38]
Subfamily Perognathinae		
Genus *Chaetodipus*	Coarse-haired Pocket Mice	[86]
C. arenarius (Nea)	Little Desert Pocket Mouse	[38]
C. artus (Nea)	Narrow-skulled Pocket Mouse	[38]
C. baileyi (Nea)	Bailey's Pocket Mouse	[114]

C. *californicus* (Nea)	California Pocket Mouse	[114]
C. *fallax* (Nea)	San Diego Pocket Mouse	[114]
C. *formosus* (Nea)	Long-tailed Pocket Mouse	[114]
C. *goldmani* (Nea)	Goldman's Pocket Mouse	[38]
C. *hispidus* (Nea)	Hispid Pocket Mouse	[114]
C. *intermedius* (Nea)	Rock Pocket Mouse	[114]
C. *lineatus* (Nea)	Lined Pocket Mouse	[38]
C. *nelsoni* (Nea)	Nelson's Pocket Mouse	[114]
C. *penicillatus* (Nea)	Desert Pocket Mouse	[114]
C. *pernix* (Nea)	Sinaloan Pocket Mouse	[38]
C. *spinatus* (Nea)	Spiny Pocket Mouse	[114]
Genus *Perognathus*	Silky Pocket Mice	[86]
P. *alticola* (Nea)	White-eared Pocket Mouse	[114]
P. *amplus* (Nea)	Arizona Pocket Mouse	[114]
P. *fasciatus* (Nea)	Olive-backed Pocket Mouse	[54]
P. *flavescens* (Nea)	Plains Pocket Mouse	[54]
P. *flavus* (Nea)	Silky Pocket Mouse	[54]
P. *inornatus* (Nea)	San Joaquin Pocket Mouse	[54]
P. *longimembris* (Nea)	Little Pocket Mouse	[114]
P. *merriami* (Nea)	Merriam's Pocket Mouse	[114]
P. *parvus* (Nea)	Great Basin Pocket Mouse	[114]
P. *xanthanotus* (Nea)	Yellow-eared Pocket Mouse	[54]
Family Dipodidae	Jerboas	[16]
Subfamily Allactaginae		
Genus *Allactaga*	Four- and Five-toed Jerboas	[86]
A. *balikunica* (P)	Balikun Jerboa	
A. *bullata* (P)	Gobi Jerboa	[109]
A. *elater* (P)	Small Five-toed Jerboa	[16]
A. *euphratica* (P)	Euphrates Jerboa	[16]
A. *firouzi* (P)	Iranian Jerboa	[109]
A. *hotsoni* (P)	Hotson's Jerboa	[16]
A. *major* (P)	Great Jerboa	[16]
A. *severtzovi* (P)	Severtzov's Jerboa	[16]
A. *sibirica* (P)	Mongolian Five-toed Jerboa	[16]
A. *tetradactyla* (P)	Four-toed Jerboa	[16]
A. *vinogradovi* (P)	Vinogradov's Jerboa	[109]
Genus *Allactodipus*		
A. *bobrinskii* (P)	Bobrinski's Jerboa	[16]
Genus *Pygeretmus*	Fat-tailed Jerboas	[16]
P. *platyurus* (P)	Lesser Fat-tailed Jerboa	[16]

P. pumilio (P)	Dwarf Fat-tailed Jerboa	
P. shitkovi (P)	Greater Fat-tailed Jerboa	[16]
Subfamily Cardiocraniinae		
Genus *Cardiocranius*		
C. paradoxus (P)	Five-toed Pygmy Jerboa	[16]
Genus *Salpingotus*	Three-toed Pygmy Jerboas	[16]
S. crassicauda (P)	Thick-tailed Pygmy Jerboa	[16]
S. heptneri (P)	Heptner's Pygmy Jerboa	[16]
S. kozlovi (P)	Kozlov's Pygmy Jerboa	[16]
S. michaelis (P)	Baluchistan Pygmy Jerboa	[16]
S. pallidus (P)	Pallid Pygmy Jerboa	
S. thomasi (P)	Thomas's Pygmy Jerboa	[109]
Subfamily Dipodinae		
Genus *Dipus*		
D. sagitta (P)	Northern Three-toed Jerboa	[16]
Genus *Eremodipus*		
E. lichtensteini (P)	Lichtenstein's Jerboa	[16]
Genus *Jaculus*	Desert Jerboas	[86]
J. blanfordi (P)	Blanford's Jerboa	[16]
J. jaculus (P, E)	Lesser Egyptian Jerboa	[16]
J. orientalis (P)	Greater Egyptian Jerboa	[16]
J. turcmenicus (P)	Turkmen Jerboa	[16]
Genus *Stylodipus*	Three-toed Jerboas	
S. andrewsi (P)	Andrews's Three-toed Jerboa	
S. sungorus (P)	Mongolian Three-toed Jerboa	
S. telum (P)	Thick-tailed Three-toed Jerboa	[16]
Subfamily Euchoreutinae		
Genus *Euchoreutes*		
E. naso (P)	Long-eared Jerboa	[16]
Subfamily Paradipodinae		
Genus *Paradipus*		
P. ctenodactylus (P)	Comb-toed Jerboa	[16]
Subfamily Sicistinae		
Genus *Sicista*	Birch Mice	[16]
S. armenica (P)	Armenian Birch Mouse	
S. betulina (P)	Northern Birch Mouse	[16]
S. caucasica (P)	Caucasian Birch Mouse	

S. caudata (P)	Long-tailed Birch Mouse	[109]
S. concolor (P, Or)	Chinese Birch Mouse	[109]
S. kazbegica (P)	Kazbeg Birch Mouse	
S. kluchorica (P)	Kluchor Birch Mouse	[109]
S. napaea (P)	Altai Birch Mouse	[16]
S. pseudonapaea (P)	Gray Birch Mouse	[109]
S. severtzovi (P)	Severtzov's Birch Mouse	
S. strandi (P)	Strand's Birch Mouse	
S. subtilis (P)	Southern Birch Mouse	[16]
S. tianshanica (P)	Tien Shan Birch Mouse	[16]
Subfamily Zapodinae		
Genus *Eozapus*		
E. setchuanus (P)	Chinese Jumping Mouse	[109]
Genus *Napaeozapus*		
N. insignis (Nea)	Woodland Jumping Mouse	[114]
Genus *Zapus*	Jumping Mice	[86]
Z. hudsonius (Nea)	Meadow Jumping Mouse	[114]
Z. princeps (Nea)	Western Jumping Mouse	[114]
Z. trinotatus (Nea)	Pacific Jumping Mouse	[114]
Family Muridae	Rats, Mice, Voles, Gerbils, Hamsters, and Lemmings	[86]
Subfamily Arvicolinae		
Genus *Alticola*	Mountain Voles	[109]
A. albicauda (Or)	White-tailed Mountain Vole	
A. argentatus (P, Or)	Silver Mountain Vole	[87]
A. barakshin (P)	Gobi Altai Mountain Vole	[87]
A. lemminus (P)	Lemming Vole	[87]
A. macrotis (P)	Large-eared Vole	[16]
A. montosa (Or)	Central Kashmir Vole	[10]
A. roylei (Or)	Royle's Mountain Vole	[16]
A. semicanus (P)	Mongolian Silver Vole	[87]
A. stoliczkanus (P, Or)	Stoliczka's Mountain Vole	[16]
A. stracheyi (Or)	Strachey's Mountain Vole	[87]
A. strelzowi (P)	Flat-headed Vole	[16]
A. tuvinicus (P)	Tuva Silver Vole	[87]
Genus *Arborimus*	Tree Voles	[86]
A. albipes (Nea)	White-footed Vole	[114]
A. longicaudus (Nea)	Red Tree Vole	[114]
A. pomo (Nea)	Sonoma Tree Vole	[114]

Genus *Arvicola*	Water Voles	[16]
A. sapidus (P)	Southwestern Water Vole	[16]
A. terrestris (P)	European Water Vole	[16]
Genus *Blanfordimys*	Afghan Voles	[86]
B. afghanus (P)	Afghan Vole	[16]
B. bucharicus (P)	Bucharian Vole	[87]
Genus *Chionomys*	Snow Voles	[86]
C. gud (P)	Caucasian Snow Vole	[36]
C. nivalis (P)	European Snow Vole	[36]
C. roberti (P)	Robert's Snow Vole	[36]
Genus *Clethrionomys*	Red-backed Voles	[16]
C. californicus (Nea)	Western Red-backed Vole	[114]
C. centralis (P)	Tien Shan Red-backed Vole	
C. gapperi (Nea)	Southern Red-backed Vole	[114]
C. glareolus (P)	Bank Vole	[16]
C. rufocanus (P)	Gray Red-backed Vole	[16]
C. rutilus (Nea, P)	Northern Red-backed Vole	[114]
C. sikotanensis (P)	Shikotan Vole	[36]
Genus *Dicrostonyx*	Collared Lemmings	[16]
D. exsul (Nea)	St. Lawrence Island Collared Lemming	[54]
D. groenlandicus (Nea)	Northern Collared Lemming	[114]
D. hudsonius (Nea)	Ungava Collared Lemming	[114]
D. kilangmiutak (Nea)	Victoria Collared Lemming	[54]
D. nelsoni (Nea)	Nelson's Collared Lemming	[54]
D. nunatakensis (Nea)	Ogilvie Mountain Collared Lemming	[54]
D. richardsoni (Nea)	Richardson's Collared Lemming	[114]
D. rubricatus (Nea)	Bering Collared Lemming	[54]
D. torquatus (P)	Arctic Lemming	[25]
D. unalascensis (Nea)	Unalaska Collared Lemming	[54]
D. vinogradovi (P)	Wrangel Lemming	[16]
Genus *Dinaromys*		
D. bogdanovi (P)	Martino's Snow Vole	[16]
Genus *Ellobius*	Mole Voles	[109]
E. alaicus (P)	Alai Mole Vole	[109]
E. fuscocapillus (P)	Southern Mole Vole	[109]

E. lutescens (P)	Transcaucasian Mole Vole	[87]
E. talpinus (P)	Northern Mole Vole	[109]
E. tancrei (P)	Zaisan Mole Vole	[87]
Genus Eolagurus	Yellow Steppe Lemmings	[109]
E. luteus (P)	Yellow Steppe Lemming	[109]
E. przewalskii (P)	Przewalski's Steppe Lemming	[109]
Genus Eothenomys	South Asian Voles	[109]
E. chinensis (P)	Pratt's Vole	[109]
E. custos (P)	Southwest China Vole	[116]
E. eva (P)	Ganzu Vole	[116]
E. inez (P)	Kolan Vole	[116]
E. melanogaster (P, Or)	Père David's Vole	[116]
E. olitor (P)	Chaotung Vole	[116]
E. proditor (P)	Yulungshan Vole	[116]
E. regulus (P)	Royal Vole	
E. shanseius (P)	Shansei Vole	[116]
Genus Hyperacrius	Kashmir Voles	[86]
H. fertilis (P, Or)	True's Vole	[16]
H. wynnei (P)	Murree Vole	[16]
Genus Lagurus		
L. lagurus (P)	Steppe Lemming	[16]
Genus Lasiopodomys	Brandt's Voles	[36]
L. brandtii (P)	Brandt's Vole	[36]
L. fuscus (P)	Plateau Vole	[51]
L. mandarinus (P)	Mandarin Vole	[16]
Genus Lemmiscus		
L. curtatus (Nea)	Sagebrush Vole	[114]
Genus Lemmus	Brown Lemmings	[16]
L. amurensis (P)	Amur Lemming	[16]
L. lemmus (P)	Norway Lemming	[16]
L. sibiricus (Nea, P)	Brown Lemming	[114]
Genus Microtus	Meadow Voles	[16]
M. abbreviatus (Nea)	Insular Vole	[114]
M. agrestis (P)	Field Vole	[16]
M. arvalis (P)	Common Vole	[16]
M. bavaricus (P)	Bavarian Pine Vole	[16]
M. breweri (Nea)	Beach Vole	[114]
M. cabrerae (P)	Cabrera's Vole	[16]
M. californicus (Nea)	California Vole	[114]

M. canicaudus (Nea)	Gray-tailed Vole	[114]
M. chrotorrhinus (Nea)	Rock Vole	[114]
M. daghestanicus (P)	Daghestan Pine Vole	[16]
M. duodecimcostatus (P)	Mediterranean Pine Vole	[16]
M. evoronensis (P)	Evorsk Vole	[109]
M. felteni (P)	Felten's Vole	
M. fortis (P)	Reed Vole	[16]
M. gerbei (P)	Gerbe's Vole	
M. gregalis (P)	Narrow-headed Vole	[16]
M. guatemalensis (Nea, Neo)	Guatemalan Vole	[38]
M. guentheri (P)	Günther's Vole	[16]
M. hyperboreus (P)	North Siberian Vole	[87]
M. irani (P)	Persian Vole	[16]
M. irene (P, Or)	Chinese Scrub Vole	[116]
M. juldaschi (P, Or)	Juniper Vole	[16]
M. kermanensis (P)	Baluchistan Vole	[16]
M. kirgisorum (P)	Tien Shan Vole	[16]
M. leucurus (P, Or)	Blyth's Vole	[16]
M. limnophilus (P)	Lacustrine Vole	
M. longicaudus (Nea)	Long-tailed Vole	[114]
M. lusitanicus (P)	Lusitanian Pine Vole	[16]
M. majori (P)	Major's Pine Vole	[16]
M. maximowiczii (P)	Maximowicz's Vole	[116]
M. mexicanus (Nea)	Mexican Vole	[54]
M. middendorffi (P)	Middendorf's Vole	[16]
M. miurus (Nea)	Singing Vole	[114]
M. mongolicus (P)	Mongolian Vole	[16]
M. montanus (Nea)	Montane Vole	[114]
M. montebelli (P)	Japanese Grass Vole	[16]
M. mujanensis (P)	Muisk Vole	
M. multiplex (P)	Alpine Pine Vole	[16]
M. nasarovi (P)	Nasarov's Vole	
M. oaxacensis (Nea)	Tarabundi Vole	[16]
M. obscurus (P)	Altai Vole	
M. ochrogaster (Nea)	Prairie Vole	[114]
M. oeconomus (Nea, P)	Tundra Vole	[114]
M. oregoni (Nea)	Creeping Vole	[114]
M. pennsylvanicus (Nea)	Meadow Vole	[114]
M. pinetorum (Nea)	Woodland Vole	[114]

M. quasiater (Nea)	Jalapan Pine Vole	[38]
M. richardsoni (Nea)	Water Vole	[114]
M. rossiaemeridionalis (P)	Southern Vole	
M. sachalinensis (P)	Sakhalin Vole	[16]
M. savii (P)	Savi's Pine Vole	[16]
M. schelkovnikovi (P)	Schelkovnikov's Pine Vole	[16]
M. sikimensis (P, Or)	Sikkim Vole	[16]
M. socialis (P)	Social Vole	[16]
M. subterraneus (P)	European Pine Vole	[16]
M. tatricus (P)	Tatra Pine Vole	[16]
M. thomasi (P)	Thomas's Pine Vole	[16]
M. townsendii (Nea)	Townsend's Vole	[114]
M. transcaspicus (P)	Transcaspian Vole	[16]
M. umbrosus (Nea)	Zempoaltepec Vole	[38]
M. xanthognathus (Nea)	Taiga Vole	[114]
Genus *Myopus*		
M. schisticolor (P)	Wood Lemming	[16]
Genus *Neofiber*		
N. alleni (Nea)	Round-tailed Muskrat	[114]
Genus *Ondatra*		
O. zibethicus (Nea)	Muskrat	[114]
Genus *Phaulomys*	Japanese Voles	
P. andersoni (P)	Japanese Red-backed Vole	[16]
P. smithii (P)	Smith's Vole	
Genus *Phenacomys*	Heather Voles	[109]
P. intermedius (Nea)	Western Heather Vole	[114]
P. ungava (Nea)	Eastern Heather Vole	[114]
Genus *Proedromys*		
P. bedfordi (P)	Duke of Bedford's Vole	[16]
Genus *Prometheomys*		
P. schaposchnikowi (P)	Long-clawed Mole Vole	[16]
Genus *Synaptomys*	Bog Lemmings	[16]
S. borealis (Nea)	Northern Bog Lemming	[114]
S. cooperi (Nea)	Southern Bog Lemming	[114]
Genus *Volemys*	Musser's Voles	
V. clarkei (P, Or)	Clarke's Vole	[16]
V. kikuchii (P)	Taiwan Vole	[16]
V. millicens (P)	Szechuan Vole	[16]
V. musseri (P)	Marie's Vole	

Subfamily Calomyscinae
Genus *Calomyscus* — Mouse-like Hamsters [16]
   C. bailwardi (P) — Mouse-like Hamster [16]
   C. baluchi (P) — Baluchi Mouse-like Hamster
   C. hotsoni (P) — Hotson's Mouse-like Hamster
   C. mystax (P) — Afghan Mouse-like Hamster [109]
   C. tsolovi (P) — Tsolov's Mouse-like Hamster
   C. urartensis (P) — Urartsk Mouse-like Hamster [109]

Subfamily Cricetinae
Genus *Allocricetulus* — Mongolian Hamsters
   A. curtatus (P) — Mongolian Hamster [16]
   A. eversmanni (P) — Eversmann's Hamster [16]
Genus *Cansumys*
   C. canus (P) — Gansu Hamster [16]
Genus *Cricetulus* — Dwarf Hamsters [16]
   C. alticola (P, Or) — Tibetan Dwarf Hamster [109]
   C. barabensis (P) — Striped Dwarf Hamster [109]
   C. kamensis (P) — Kam Dwarf Hamster [109]
   C. longicaudatus (P, Or) — Long-tailed Dwarf Hamster [109]
   C. migratorius (P) — Gray Dwarf Hamster [109]
   C. sokolovi (P) — Sokolov's Dwarf Hamster
Genus *Cricetus*
   C. cricetus (P) — Black-bellied Hamster [86]
Genus *Mesocricetus* — Golden Hamsters [86]
   M. auratus (P) — Golden Hamster [16]
   M. brandti (P) — Brandt's Hamster [109]
   M. newtoni (P) — Romanian Hamster [109]
   M. raddei (P) — Ciscaucasian Hamster [16]
Genus *Phodopus* — Small Desert Hamsters [86]
   P. campbelli (P) — Campbell's Hamster
   P. roborovskii (P) — Desert Hamster [16]
   P. sungorus (P) — Dzhungarian Hamster [109]
Genus *Tscherskia*
   T. triton (P) — Greater Long-tailed Hamster [16]

Subfamily Cricetomyinae

Genus *Beamys*	Long-tailed Pouched Rats	[16]
B. *hindei* (E)	Long-tailed Pouched Rat	[16]
B. *major* (E)	Greater Long-tailed Pouched Rat	
Genus *Cricetomys*	African Giant Pouched Rats	[86]
C. *emini* (E)	Giant Rat	[22]
C. *gambianus* (E)	Gambian Rat	[109]
Genus *Saccostomus*	Pouched Mice	[63]
S. *campestris* (E)	Pouched Mouse	[16]
S. *mearnsi* (E)	Mearns's Pouched Mouse	

Subfamily Dendromurinae

Genus *Dendromus*	African Climbing Mice	[16]
D. *insignis* (E)	Remarkable Climbing Mouse	
D. *kahuziensis* (E)	Mt. Kahuzi Climbing Mouse	
D. *kivu* (E)	Kivu Climbing Mouse	
D. *lovati* (E)	Lovat's Climbing Mouse	
D. *melanotis* (E)	Gray Climbing Mouse	[16]
D. *mesomelas* (E)	Brant's Climbing Mouse	[16]
D. *messorius* (E)	Banana Climbing Mouse	
D. *mystacalis* (E)	Chestnut Climbing Mouse	[16]
D. *nyikae* (E)	Nyika Climbing Mouse	[16]
D. *oreas* (E)	Cameroon Climbing Mouse	
D. *vernayi* (E)	Vernay's Climbing Mouse	
Genus *Dendroprionomys*		
D. *rousseloti* (E)	Velvet Climbing Mouse	[63]
Genus *Deomys*		
D. *ferrugineus* (E)	Congo Forest Mouse	[16]
Genus *Leimacomys*		
L. *buettneri* (E)	Groove-toothed Forest Mouse	[109]
Genus *Malacothrix*		
M. *typica* (E)	Gerbil Mouse	[16]
Genus *Megadendromus*		
M. *nikolausi* (E)	Nikolaus's Mouse	
Genus *Prionomys*		
P. *batesi* (E)	Dollman's Tree Mouse	[16]

Genus *Steatomys*	Fat Mice	[16]
S. *caurinus* (E)	Northwestern Fat Mouse	[109]
S. *cuppedius* (E)	Dainty Fat Mouse	[109]
S. *jacksoni* (E)	Jackson's Fat Mouse	[109]
S. *krebsii* (E)	Kreb's Fat Mouse	[16]
S. *parvus* (E)	Tiny Fat Mouse	[16]
S. *pratensis* (E)	Fat Mouse	[107]
Subfamily Gerbillinae		
Genus *Ammodillus*		
A. *imbellis* (E)	Ammodile	[109]
Genus *Brachiones*		
B. *przewalskii* (P)	Przewalski's Gerbil	[16]
Genus *Desmodilliscus*		
D. *braueri* (E)	Pouched Gerbil	[16]
Genus *Desmodillus*		
D. *auricularis* (E)	Cape Short-eared Gerbil	[16]
Genus *Gerbillurus*	Hairy-footed Gerbils	
G. *paeba* (E)	Hairy-footed Gerbil	[16]
G. *setzeri* (E)	Setzer's Hairy-footed Gerbil	[16]
G. *tytonis* (E)	Dune Hairy-footed Gerbil	[107]
G. *vallinus* (E)	Bushy-tailed Hairy-footed Gerbil	[16]
Genus *Gerbillus*	Gerbils	
G. *acticola* (E)	Berbera Gerbil	
G. *agag* (E)	Agag Gerbil	[104]
G. *allenbyi* (P)	Allenby's Gerbil	
G. *amoenus* (P)	Pleasant Gerbil	
G. *andersoni* (P)	Anderson's Gerbil	[16]
G. *aquilus* (P)	Swarthy Gerbil	
G. *bilensis* (E)	Bilen Gerbil	
G. *bonhotei* (P)	Bonhote's Gerbil	
G. *bottai* (E)	Botta's Gerbil	
G. *brockmani* (E)	Brockman's Gerbil	
G. *burtoni* (E)	Burton's Gerbil	
G. *campestris* (P, E)	North African Gerbil	[109]
G. *cheesmani* (P)	Cheesman's Gerbil	[16]
G. *cosensis* (E)	Cosens's Gerbil	
G. *dalloni* (E)	Dallon's Gerbil	
G. *dasyurus* (P)	Wagner's Gerbil	[16]
G. *diminutus* (E)	Diminutive Gerbil	

G. dongolanus (E)	Dongola Gerbil	
G. dunni (E)	Somalia Gerbil	[109]
G. famulus (P)	Black-tufted Gerbil	[16]
G. floweri (P)	Flower's Gerbil	
G. garamantis (P)	Algerian Gerbil	
G. gerbillus (P, E)	Lesser Egyptian Gerbil	[109]
G. gleadowi (P, Or)	Indian Hairy-footed Gerbil	[109]
G. grobbeni (P)	Grobben's Gerbil	
G. harwoodi (E)	Harwood's Gerbil	
G. henleyi (P, E)	Pygmy Gerbil	[16]
G. hesperinus (P)	Western Gerbil	
G. hoogstraali (P)	Hoogstraal's Gerbil	
G. jamesi (P)	James's Gerbil	
G. juliani (E)	Julian's Gerbil	
G. latastei (P)	Lataste's Gerbil	
G. lowei (E)	Lowe's Gerbil	
G. mackillingini (P, E)	Mackillingin's Gerbil	[109]
G. maghrebi (P)	Greater Short-tailed Gerbil	[16]
G. mauritaniae (E)	Mauritanian Gerbil	[109]
G. mesopotamiae (P)	Mesopotamian Gerbil	[16]
G. muriculus (E)	Barfur Gerbil	[109]
G. nancillus (E)	Sudan Gerbil	[109]
G. nanus (P, Or)	Baluchistan Gerbil	[16]
G. nigeriae (E)	Nigerian Gerbil	
G. occiduus (P)	Occidental Gerbil	
G. percivali (E)	Percival's Gerbil	
G. perpallidus (P)	Pale Gerbil	
G. poecilops (P)	Large Aden Gerbil	[16]
G. principulus (E)	Principal Gerbil	
G. pulvinatus (E)	Cushioned Gerbil	[100]
G. pusillus (E)	Least Gerbil	
G. pyramidum (P, E)	Greater Egyptian Gerbil	[16]
G. quadrimaculatus (E)	Four-spotted Gerbil	
G. riggenbachi (P, E)	Riggenbach's Gerbil	[109]
G. rosalinda (E)	Rosalinda Gerbil	
G. ruberrimus (E)	Little Red Gerbil	[100]
G. simoni (P)	Lesser Short-tailed Gerbil	[16]
G. somalicus (E)	Somalian Gerbil	
G. stigmonyx (E)	Khartoum Gerbil	
G. syrticus (P)	Sand Gerbil	

G. tarabuli (P)	Tarabul's Gerbil	
G. vivax (P)	Vivacious Gerbil	
G. watersi (E)	Waters's Gerbil	[109]
Genus *Meriones*	Jirds	[16]
M. arimalius (P)	Arabian Jird	
M. chengi (P)	Cheng's Jird	
M. crassus (P)	Sundevall's Jird	[109]
M. dahli (P)	Dahl's Jird	
M. hurrianae (P, Or)	Indian Desert Jird	[109]
M. libycus (P)	Libyan Jird	[16]
M. meridianus (P)	Mid-day Jird	[109]
M. persicus (P)	Persian Jird	[16]
M. rex (P)	King Jird	[16]
M. sacramenti (P)	Buxton's Jird	[16]
M. shawi (P)	Shaw's Jird	[16]
M. tamariscinus (P)	Tamarisk Jird	[109]
M. tristrami (P)	Tristram's Jird	[16]
M. unguiculatus (P)	Mongolian Jird	[109]
M. vinogradovi (P)	Vinogradov's Jird	[16]
M. zarudnyi (P)	Zarudny's Jird	[109]
Genus *Microdillus*		
M. peeli (E)	Somali Pygmy Gerbil	[16]
Genus *Pachyuromys*		
P. duprasi (P)	Fat-tailed Gerbil	[16]
Genus *Psammomys*	Sand Rats	[109]
P. obesus (P)	Fat Sand Rat	[16]
P. vexillaris (P)	Thin Sand Rat	
Genus *Rhombomys*		
R. opimus (P)	Great Gerbil	[16]
Genus *Sekeetamys*		
S. calurus (P)	Bushy-tailed Jird	[16]
Genus *Tatera*	Large Naked-soled Gerbils	[86]
T. afra (E)	Cape Gerbil	[16]
T. boehmi (E)	Boehm's Gerbil	[16]
T. brantsii (E)	Highveld Gerbil	[16]
T. guineae (E)	Guinea Gerbil	[104]
T. inclusa (E)	Gorongoza Gerbil	[16]
T. indica (P, Or)	Indian Gerbil	[109]
T. kempi (E)	Kemp's Gerbil	[104]
T. leucogaster (E)	Bushveld Gerbil	[16]

T. nigricauda (E)	Black-tailed Gerbil	[16]
T. phillipsi (E)	Phillip's Gerbil	
T. robusta (E)	Fringe-tailed Gerbil	[16]
T. valida (E)	Savanna Gerbil	[16]
Genus *Taterillus*	Small Naked-soled Gerbils	[86]
T. arenarius (E)	Sahel Gerbil	
T. congicus (E)	Congo Gerbil	[109]
T. emini (E)	Emin's Gerbil	[109]
T. gracilis (E)	Slender Gerbil	[109]
T. harringtoni (E)	Harrington's Gerbil	[109]
T. lacustris (E)	Lake Chad Gerbil	[41]
T. petteri (E)	Petter's Gerbil	
T. pygargus (E)	Senegal Gerbil	[109]
Subfamily Lophiomyinae		
Genus *Lophiomys*		
L. imhausi (E)	Crested Rat	[16]
Subfamily Murinae		
Genus *Abditomys*		
A. latidens (Or)	Luzon Broad-toothed Rat	[48]
Genus *Acomys*	African Spiny Mice	[16]
A. cahirinus (P, E)	Cairo Spiny Mouse	[16]
A. cilicicus (P)	Asia Minor Spiny Mouse	[109]
A. cinerasceus (E)	Gray Spiny Mouse	
A. ignitus (E)	Fiery Spiny Mouse	
A. kempi (E)	Kemp's Spiny Mouse	
A. louisae (E)	Louise's Spiny Mouse	
A. minous (P)	Crete Spiny Mouse	
A. mullah (E)	Mullah Spiny Mouse	
A. nesiotes (P)	Cyprus Spiny Mouse	
A. percivali (E)	Percival's Spiny Mouse	
A. russatus (P)	Golden Spiny Mouse	[16]
A. spinosissimus (E)	Spiny Mouse	[107]
A. subspinosus (E)	Cape Spiny Mouse	[107]
A. wilsoni (E)	Wilson's Spiny Mouse	[109]
Genus *Aethomys*	African Rock Rats	
A. bocagei (E)	Bocage's Rock Rat	
A. chrysophilus (E)	Red Rock Rat	
A. granti (E)	Grant's Rock Rat	
A. hindei (E)	Hinde's Rock Rat	
A. kaiseri (E)	Kaiser's Rock Rat	

A. namaquensis (E)	Namaqua Rock Rat	
A. nyikae (E)	Nyika Rock Rat	
A. silindensis (E)	Silinda Rock Rat	
A. stannarius (E)	Tinfields Rock Rat	
A. thomasi (E)	Thomas's Rock Rat	
Genus *Anisomys*		
A. imitator (A)	Squirrel-toothed Rat	[16]
Genus *Anonymomys*		
A. mindorensis (Or)	Mindoro Rat	[20]
Genus *Apodemus*	Field Mice	[86]
A. agrarius (P)	Striped Field Mouse	[16]
A. alpicola (P)	Alpine Field Mouse	
A. argenteus (P)	Small Japanese Field Mouse	[16]
A. arianus (P)	Persian Field Mouse	
A. chevrieri (P)	Chevrier's Field Mouse	
A. draco (P, Or)	South China Field Mouse	[109]
A. flavicollis (P)	Yellow-necked Field Mouse	[109]
A. fulvipectus (P)	Yellow-breasted Field Mouse	
A. gurkha (Or)	Himalayan Field Mouse	[16]
A. hermonensis (P)	Mt. Hermon Field Mouse	
A. hyrcanicus (P)	Caucasus Field Mouse	
A. latronum (P, Or)	Sichuan Field Mouse	[109]
A. mystacinus (P)	Broad-toothed Field Mouse	[109]
A. peninsulae (P)	Korean Field Mouse	[16]
A. ponticus (P)	Black Sea Field Mouse	
A. rusiges (Or)	Kashmir Field Mouse	
A. semotus (P)	Taiwan Field Mouse	[109]
A. speciosus (P)	Large Japanese Field Mouse	[16]
A. sylvaticus (P)	Long-tailed Field Mouse	[109]
A. uralensis (P)	Ural Field Mouse	
A. wardi (P, Or)	Ward's Field Mouse	
Genus *Apomys*	Philippine Forest Mice	
A. abrae (Or)	Luzon Cordillera Forest Mouse	[48]
A. datae (Or)	Luzon Montane Forest Mouse	[48]
A. hylocoetes (Or)	Mt. Apo Forest Mouse	[73]
A. insignis (Or)	Mindanao Montane Forest Mouse	[48]

A. littoralis (Or)	Mindanao Lowland Forest Mouse	[48]
A. microdon (Or)	Small Luzon Forest Mouse	[48]
A. musculus (Or)	Least Forest Mouse	[48]
A. sacobianus (Or)	Long-nosed Luzon Forest Mouse	[48]
Genus *Archboldomys*		
A. luzonensis (Or)	Mt. Isarog Shrew Mouse	[48]
Genus *Arvicanthis*	African Grass Rats	[16]
A. abyssinicus (E)	Abyssinian Grass Rat	
A. blicki (E)	Blick's Grass Rat	
A. nairobae (E)	Nairobi Grass Rat	
A. niloticus (P, E)	African Grass Rat	[109]
A. newmanni (E)	Newmann's Grass Rat	
Genus *Bandicota*	Bandicoot Rats	[16]
B. bengalensis (P, Or)	Lesser Bandicoot Rat	[16]
B. indica (P, Or)	Greater Bandicoot Rat	[16]
B. savilei (Or)	Savile's Bandicoot Rat	
Genus *Batomys*	Hairy-tailed Rats	
B. dentatus (Or)	Large-toothed Hairy-tailed Rat	[48]
B. granti (Or)	Luzon Hairy-tailed Rat	[48]
B. salomonseni (Or)	Mindanao Hairy-tailed Rat	[48]
Genus *Berylmys*	White-toothed Rats	[16]
B. berdmorei (Or)	Small White-toothed Rat	[16]
B. bowersi (P, Or)	Bower's White-toothed Rat	
B. mackenziei (P, Or)	Kenneth's White-toothed Rat	[16]
B. manipulus (Or)	Manipur White-toothed Rat	
Genus *Bullimus*	Philippine Rats	
B. bagobus (Or)	Bagobo Rat	
B. luzonicus (Or)	Luzon Forest Rat	
Genus *Bunomys*	Hill Rats	
B. andrewsi (A)	Andrew's Hill Rat	
B. chrysocomus (A)	Yellow-haired Hill Rat	
B. coelestis (A)	Heavenly Hill Rat	
B. fratrorum (A)	Fraternal Hill Rat	
B. heinrichi (A)	Heinrich's Hill Rat	

B. penitus (A)	Inland Hill Rat	
B. prolatus (A)	Long-headed Hill Rat	
Genus *Canariomys*		
C. tamarani (P)	Canary Mouse	
Genus *Carpomys*	Luzon Tree Rats	[16]
C. melanurus (Or)	Short-footed Luzon Tree Rat	[48]
C. phaeurus (Or)	White-bellied Luzon Tree Rat	[48]
Genus *Celaenomys*		
C. silaceus (Or)	Blazed Luzon Shrew Rat	[16]
Genus *Chiromyscus*		
C. chiropus (Or)	Fea's Tree Rat	[16]
Genus *Chiropodomys*	Pencil-tailed Tree Mice	[16]
C. calamianensis (Or)	Palawan Pencil-tailed Tree Mouse	[48]
C. gliroides (P, Or)	Pencil-tailed Tree Mouse	[109]
C. karlkoopmani (Or)	Koopman's Pencil-tailed Tree Mouse	
C. major (Or)	Large Pencil-tailed Tree Mouse	[91]
C. muroides (Or)	Gray-bellied Pencil-tailed Tree Mouse	[91]
C. pusillus (Or)	Small Pencil-tailed Tree Mouse	
Genus *Chiruromys*	Tree Mice	
C. forbesi (A)	Greater Tree Mouse	[16]
C. lamia (A)	Broad-headed Tree Mouse	[28]
C. vates (A)	Lesser Tree Mouse	[16]
Genus *Chrotomys*	Philippine Striped Rats	[16]
C. gonzalesi (Or)	Isarog Striped Shrew-Rat	[48]
C. mindorensis (Or)	Mindoro Striped Rat	[16]
C. whiteheadi (Or)	Luzon Striped Rat	[16]
Genus *Coccymys*	Brush Mice	
C. albidens (A)	White-toothed Brush Mouse	
C. ruemmleri (A)	Rümmler's Brush Mouse	[16]
Genus *Colomys*		
C. goslingi (E)	African Water Rat	[109]

Genus *Conilurus*	Rabbit Rats	[86]
C. albipes (A)	White-footed Rabbit Rat	[16]
C. penicillatus (A)	Brush-tailed Rabbit Rat	[16]
Genus *Coryphomys*		
C. buhleri (A)	Buhler's Rat	
Genus *Crateromys*	Bushy-tailed Cloud Rats	[16]
C. australis (Or)	Dinagat Bushy-tailed Cloud Rat	
C. paulus (Or)	Ilin Bushy-tailed Cloud Rat	[37]
C. schadenbergi (Or)	Luzon Bushy-tailed Cloud Rat	[48]
Genus *Cremnomys*	Indian Rats	
C. blanfordi (Or)	Blanford's Rat	
C. cutchicus (Or)	Cutch Rat	[109]
C. elvira (Or)	Elvira Rat	
Genus *Crossomys*		
C. moncktoni (A)	Earless Water Rat	[16]
Genus *Crunomys*	Philippine Shrew Rats	[86]
C. celebensis (A)	Celebes Shrew Rat	
C. fallax (Or)	Northern Luzon Shrew Rat	
C. melanius (Or)	Mindanao Shrew Rat	
C. rabori (Or)	Leyte Shrew Rat	
Genus *Dacnomys*		
D. millardi (P, Or)	Millard's Rat	[16]
Genus *Dasymys*	Shaggy African Marsh Rats	[109]
D. foxi (E)	Fox's Shaggy Rat	[104]
D. incomtus (E)	African Marsh Rat	[16]
D. montanus (E)	Montane Shaggy Rat	
D. nudipes (E)	Angolan Marsh Rat	[101]
D. rufulus (E)	West African Shaggy Rat	[104]
Genus *Dephomys*	Defua Rats	[86]
D. defua (E)	Defua Rat	[104]
D. eburnea (E)	Ivory Coast Rat	
Genus *Desmomys*		
D. harringtoni (E)	Harrington's Rat	
Genus *Diomys*		
D. crumpi (Or)	Crump's Mouse	[16]
Genus *Diplothrix*		
D. legatus (P)	Ryukyu Rat	

Genus *Echiothrix*
    E. leucura (A)   Sulawesi Spiny Rat   [16]
Genus *Eropeplus*
    E. canus (A)   Sulawesi Soft-furred Rat   [16]
Genus *Golunda*
    G. ellioti (P, Or)   Indian Bush Rat   [16]
Genus *Grammomys*   African Thicket Rats   [109]
    G. aridulus (E)   Arid Thicket Rat
    G. buntingi (E)   Bunting's Thicket Rat   [104]
    G. caniceps (E)   Gray-headed Thicket Rat
    G. cometes (E)   Mozambique Thicket Rat
    G. dolichurus (E)   Woodland Thicket Rat
    G. dryas (E)   Forest Thicket Rat
    G. gigas (E)   Giant Thicket Rat
    G. ibeanus (E)   Ruwenzori Thicket Rat
    G. macmillani (E)   Macmillan's Thicket Rat
    G. minnae (E)   Ethiopian Thicket Rat
    G. rutilans (E)   Shining Thicket Rat   [109]
Genus *Hadromys*
    H. humei (P, Or)   Manipur Bush Rat   [16]
Genus *Haeromys*   Ranee Mice
    H. margarettae (Or)   Ranee Mouse   [16]
    H. minahassae (A)   Minahassa Ranee Mouse
    H. pusillus (Or)   Lesser Ranee Mouse   [16]
Genus *Hapalomys*   Marmoset Rats   [16]
    H. delacouri (P, Or)   Delacour's Marmoset Rat
    H. longicaudatus (Or)   Marmoset Rat   [109]
Genus *Heimyscus*
    H. fumosus (E)   African Smoky Mouse
Genus *Hybomys*   Striped Mice
    H. basilii (E)   Father Basilio's Striped Mouse
    H. eisentrauti (E)   Eisentraut's Striped Mouse
    H. lunaris (E)   Moon Striped Mouse
    H. planifrons (E)   Miller's Striped Mouse
    H. trivirgatus (E)   Temminck's Striped Mouse   [109]
    H. univittatus (E)   Peter's Striped Mouse   [109]
Genus *Hydromys*   Water Rats   [86]
    H. chrysogaster (A)   Golden-bellied Water Rat
    H. habbema (A)   Mountain Water Rat   [16]

H. hussoni (A)	Western Water Rat	[28]
H. neobrittanicus (A)	New Britain Water Rat	
H. shawmayeri (A)	Shaw Mayer's Water Rat	
Genus Hylomyscus	African Wood Mice	[16]
H. aeta (E)	Beaded Wood Mouse	[104]
H. alleni (E)	Allen's Wood Mouse	[104]
H. baeri (E)	Baer's Wood Mouse	[104]
H. carillus (E)	Angolan Wood Mouse	
H. denniae (E)	Montane Wood Mouse	
H. parvus (E)	Little Wood Mouse	
H. stella (E)	Stella Wood Mouse	
Genus Hyomys	New Guinean Giant Rats	[109]
H. dammermani (A)	Western White-eared Giant Rat	[28]
H. goliath (A)	Eastern White-eared Giant Rat	[28]
Genus Kadarsanomys		
K. sodyi (Or)	Sody's Tree Rat	[20]
Genus Komodomys		
K. rintjanus (A)	Komodo Rat	[16]
Genus Lamottemys		
L. okuensis (E)	Mt. Oku Rat	
Genus Leggadina	Australian Native Mice	[109]
L. forresti (A)	Forrest's Mouse	[110]
L. lakedownensis (A)	Lakeland Downs Mouse	[16]
Genus Lemniscomys	Striped Grass Mice	[86]
L. barbarus (P, E)	Barbary Striped Grass Mouse	
L. bellieri (E)	Bellier's Striped Grass Mouse	
L. griselda (E)	Griselda's Striped Grass Mouse	
L. hoogstraali (E)	Hoogstral's Striped Grass Mouse	
L. linulus (E)	Senegal One-striped Grass Mouse	[104]
L. macculus (E)	Buffoon Striped Grass Mouse	
L. mittendorfi (E)	Mittendorf's Striped Grass Mouse	

L. rosalia (E)	Single-striped Grass Mouse	
L. roseveari (E)	Rosevear's Striped Grass Mouse	
L. striatus (E)	Typical Striped Grass Mouse	
Genus *Lenomys*		
L. meyeri (A)	Trefoil-toothed Giant Rat	[86]
Genus *Lenothrix*		
L. canus (Or)	Gray Tree Rat	[16]
Genus *Leopoldamys*	Long-tailed Giant Rats	[86]
L. edwardsi (P, Or)	Edwards's Long-tailed Giant Rat	
L. neilli (Or)	Neill's Long-tailed Giant Rat	
L. sabanus (Or)	Long-tailed Giant Rat	[16]
L. siporanus (Or)	Mentawai Long-tailed Giant Rat	[20]
Genus *Leporillus*	Australian Stick-nest Rats	[86]
L. apicalis (A)	Lesser Stick-nest Rat	[110]
L. conditor (A)	Greater Stick-nest Rat	[110]
Genus *Leptomys*	New Guinean Water Rats	[109]
L. elegans (A)	Long-footed Water Rat	
L. ernstmayri (A)	Ernst Mayr's Water Rat	
L. signatus (A)	Fly River Water Rat	
Genus *Limnomys*		
L. sibuanus (Or)	Mindanao Mountain Rat	[20]
Genus *Lophuromys*	Brush-furred Rats	[16]
L. cinereus (E)	Gray Brush-furred Rat	
L. flavopunctatus (E)	Yellow-spotted Brush-furred Rat	
L. luteogaster (E)	Yellow-bellied Brush-furred Rat	
L. medicaudatus (E)	Medium-tailed Brush-furred Rat	
L. melanonyx (E)	Black-clawed Brush-furred Rat	
L. nudicaudus (E)	Fire-bellied Brush-furred Rat	
L. rahmi (E)	Rahm's Brush-furred Rat	

L. sikapusi (E)	Rusty-bellied Brush-furred Rat	
L. woosnami (E)	Woosnam's Brush-furred Rat	
Genus *Lorentzimys*		
L. nouhuysi (A)	New Guinean Jumping Mouse	[16]
Genus *Macruromys*	New Guinean Rats	[86]
M. elegans (A)	Western Small-toothed Rat	[16]
M. major (A)	Eastern Small-toothed Rat	[16]
Genus *Malacomys*	African Swamp Rats	[16]
M. cansdalei (E)	Cansdale's Swamp Rat	
M. edwardsi (E)	Edward's Swamp Rat	[41]
M. longipes (E)	Big-eared Swamp Rat	[109]
M. lukolelae (E)	Lukolela Swamp Rat	
M. verschureni (E)	Verschuren's Swamp Rat	
Genus *Mallomys*	Woolly Rats	
M. aroaensis (A)	De Vis's Woolly Rat	[28]
M. gunung (A)	Alpine Woolly Rat	[28]
M. istapantap (A)	Subalpine Woolly Rat	[28]
M. rothschildi (A)	Rothschild's Woollly Rat	[28]
Genus *Malpaisomys*		
M. insularis (P)	Lava Mouse	
Genus *Margaretamys*	Margareta's Rats	[86]
M. beccarii (A)	Beccari's Margareta Rat	
M. elegans (A)	Elegant Margareta Rat	
M. parvus (A)	Little Margareta Rat	
Genus *Mastomys*	Multimammate Mice	
M. angolensis (E)	Angolan Multimammate Mouse	
M. coucha (E)	Southern Multimammate Mouse	
M. erythroleucus (P, E)	Guinea Multimammate Mouse	
M. hildebrandtii (E)	Hildebrandt's Multimammate Mouse	
M. natalensis (E)	Natal Multimammate Mouse	[16]
M. pernanus (E)	Dwarf Multimammate Mouse	

M. shortridgei (E)	Shortridge's Multimammate Mouse	
M. verheyeni (E)	Verhayen's Multimammate Mouse	
Genus *Maxomys*	Oriental Spiny Rats	[16]
M. alticola (Or)	Mountain Spiny Rat	[16]
M. baeodon (Or)	Small Spiny Rat	[16]
M. bartelsii (Or)	Bartels's Spiny Rat	
M. dollmani (A)	Dollman's Spiny Rat	
M. hellwaldii (A)	Hellwald's Spiny Rat	
M. hylomyoides (Or)	Sumatran Spiny Rat	
M. inas (Or)	Malayan Mountain Spiny Rat	[16]
M. inflatus (Or)	Fat-nosed Spiny Rat	
M. moi (Or)	Mo's Spiny Rat	
M. musschenbroekii (A)	Musschenbroek's Spiny Rat	
M. ochraceiventer (Or)	Chestnut-bellied Spiny Rat	[16]
M. pagensis (Or)	Pagai Spiny Rat	
M. panglima (Or)	Palawan Spiny Rat	[48]
M. rajah (Or)	Rajah Spiny Rat	
M. surifer (Or)	Red Spiny Rat	[16]
M. wattsi (A)	Watts's Spiny Rat	
M. whiteheadi (Or)	Whitehead's Spiny Rat	
Genus *Mayermys*		
M. ellermani (A)	One-toothed Shrew Mouse	[28]
Genus *Melasmothrix*		
M. naso (A)	Sulawesian Shrew Rat	
Genus *Melomys*	Mosaic-tailed Rats	[16]
M. aerosus (A)	Dusky Mosaic-tailed Rat	
M. bougainville (A)	Bougainville Mosaic-tailed Rat	
M. burtoni (A)	Grassland Mosaic-tailed Rat	
M. capensis (A)	Cape York Mosaic-tailed Rat	
M. cervinipes (A)	Fawn-footed Mosaic-tailed Rat	
M. fellowsi (A)	Red-bellied Mosaic-tailed Rat	
M. fraterculus (A)	Manusela Mosaic-tailed Rat	
M. gracilis (A)	Slender Mosaic-tailed Rat	

M. lanosus (A)	Large-scaled Mosaic-tailed Rat	
M. leucogaster (A)	White-bellied Mosaic-tailed Rat	
M. levipes (A)	Long-nosed Mosaic-tailed Rat	
M. lorentzii (A)	Lorentz's Mosaic-tailed Rat	
M. mollis (A)	Thomas's Mosaic-tailed Rat	
M. moncktoni (A)	Moncton's Mosaic-tailed Rat	
M. obiensis (A)	Obi Mosaic-tailed Rat	
M. platyops (A)	Lowland Mosaic-tailed Rat	
M. rattoides (A)	Large Mosaic-tailed Rat	
M. rubex (A)	Mountain Mosaic-tailed Rat	
M. rubicola (A)	Bramble Cay Mosaic-tailed Rat	
M. rufescens (A)	Black-tailed Mosaic-tailed Rat	
M. spechti (A)	Specht's Mosaic-tailed Rat	
Genus *Mesembriomys*	Tree Rats	[86]
M. gouldii (A)	Black-footed Tree Rat	[110]
M. macrurus (A)	Golden-backed Tree Rat	[110]
Genus *Microhydromys*	Lesser Shrew Mice	
M. musseri (A)	Musser's Shrew Mouse	[28]
M. richardsoni (A)	Groove-toothed Shrew Mouse	[28]
Genus *Micromys*		
M. minutus (P, Or)	Eurasian Harvest Mouse	[109]
Genus *Millardia*	Asian Soft-furred Rats	[86]
M. gleadowi (P, Or)	Sand-colored Soft-furred Rat	
M. kathleenae (Or)	Miss Ryley's Soft-furred Rat	
M. kondana (Or)	Kondana Soft-furred Rat	
M. meltada (P, Or)	Soft-furred Rat	
Genus *Muriculus*		
M. imberbis (E)	Striped-back Mouse	[109]
Genus *Mus*	Old World Mice	[86]
M. baoulei (E)	Baoule's Mouse	
M. booduga (Or)	Little Indian Field Mouse	[109]
M. bufo (E)	Toad Mouse	

M. callewaerti (E)	Callewaert's Mouse	
M. caroli (P, Or)	Ryukyu Mouse	[16]
M. cervicolor (Or)	Fawn-colored Mouse	[16]
M. cookii (P, Or)	Cook's Mouse	[16]
M. crociduroides (Or)	Sumatran Shrewlike Mouse	
M. famulus (Or)	Servant Mouse	
M. fernandoni (Or)	Ceylon Spiny Mouse	[93]
M. goundae (E)	Gounda Mouse	
M. haussa (E)	Hausa Mouse	[109]
M. indutus (E)	Desert Pygmy Mouse	[16]
M. kasaicus (E)	Kasai Mouse	
M. macedonicus (P)	Macedonian Mouse	
M. mahomet (E)	Mahomet Mouse	
M. mattheyi (E)	Matthey's Mouse	
M. mayori (Or)	Mayor's Mouse	[109]
M. minutoides (E)	Pygmy Mouse	[16]
M. musculoides (E)	Temminck's Mouse	[104]
M. musculus (All)	House Mouse	[16]
M. neavei (E)	Neave's Mouse	
M. orangiae (E)	Orange Mouse	
M. oubanguii (E)	Oubangui Mouse	
M. pahari (P, Or)	Gairdner's Shrewmouse	[16]
M. phillipsi (Or)	Phillips's Mouse	
M. platythrix (Or)	Flat-haired Mouse	[109]
M. saxicola (P, Or)	Rock-loving Mouse	
M. setulosus (E)	Peter's Mouse	[104]
M. setzeri (E)	Setzer's Pygmy Mouse	[16]
M. shortridgei (Or)	Shortridge's Mouse	[16]
M. sorella (E)	Thomas's Pygmy Mouse	[107]
M. spicilegus (P)	Mound-building Mouse	[71]
M. spretus (P)	Algerian Mouse	[16]
M. tenellus (E)	Delicate Mouse	
M. terricolor (P, Or)	Earth-colored Mouse	
M. triton (E)	Gray-bellied Pygmy Mouse	[16]
M. vulcani (Or)	Volcano Mouse	
Genus *Mylomys*		
M. dybowskii (E)	African Groove-toothed Rat	[86]
Genus *Myomys*	African Mice	
M. albipes (E)	Ethiopian White-footed Mouse	

M. daltoni (E)	Dalton's Mouse	[109]
M. derooi (E)	Deroo's Mouse	[41]
M. fumatus (E)	African Rock Mouse	
M. ruppi (E)	Rupp's Mouse	
M. verreauxii (E)	Verreaux's Mouse	[16]
M. yemeni (P)	Yemeni Mouse	
Genus *Neohydromys*		
N. fuscus (A)	Mottled-tailed Shrew Mouse	[28]
Genus *Nesokia*	Short-tailed Bandicoot Rats	[109]
N. bunnii (P)	Bunn's Short-tailed Bandicoot Rat	
N. indica (P, Or)	Short-tailed Bandicoot Rat	[16]
Genus *Niviventer*	White-bellied Rats	[86]
N. andersoni (P)	Anderson's White-bellied Rat	
N. brahma (Or)	Brahma White-bellied Rat	
N. confucianus (P, Or)	Chinese White-bellied Rat	[16]
N. coxingi (P)	Coxing's White-bellied Rat	
N. cremoriventer (Or)	Dark-tailed Tree Rat	[16]
N. culturatus (P)	Oldfield White-bellied Rat	
N. eha (P, Or)	Smoke-bellied Rat	[16]
N. excelsior (P)	Large White-bellied Rat	
N. fulvescens (P, Or)	Chestnut White-bellied Rat	
N. hinpoon (Or)	Limestone Rat	[67]
N. langbianis (Or)	Lang Bian White-bellied Rat	
N. lepturus (Or)	Narrow-tailed White-bellied Rat	
N. niviventer (P, Or)	White-bellied Rat	[16]
N. rapit (Or)	Long-tailed Mountain Rat	[16]
N. tenaster (Or)	Tenasserim White-bellied Rat	
Genus *Notomys*	Australian Hopping Mice	[16]
N. alexis (A)	Spinifex Hopping Mouse	[110]
N. amplus (A)	Short-tailed Hopping Mouse	[110]
N. aquilo (A)	Northern Hopping Mouse	[110]
N. cervinus (A)	Fawn Hopping Mouse	[110]
N. fuscus (A)	Dusky Hopping Mouse	[110]

N. longicaudatus (A)	Long-tailed Hopping Mouse	[110]
N. macrotis (A)	Big-eared Hopping Mouse	[110]
N. mitchellii (A)	Mitchell's Hopping Mouse	[110]
N. mordax (A)	Darling Downs Hopping Mouse	[110]
Genus *Oenomys*	Rufous-nosed Rats	[109]
O. hypoxanthus (E)	Rufous-nosed Rat	[16]
O. ornatus (E)	Ghana Rufous-nosed Rat	[104]
Genus *Palawanomys*		
P. furvus (Or)	Palawan Soft-furred Mountain Rat	[48]
Genus *Papagomys*	Flores Island Giant Tree Rats	[109]
P. armandvillei (A)	Flores Giant Tree Rat	
P. theodorverhoeveni (A)	Verhoeven's Giant Tree Rat	
Genus *Parahydromys*		
P. asper (A)	Coarse-haired Water Rat	
Genus *Paraleptomys*	Montane Water Rats	
P. rufilatus (A)	Northern Water Rat	
P. wilhelmina (A)	Short-haired Water Rat	
Genus *Paruromys*	Sulawesi Giant Rats	
P. dominator (A)	Sulawesi Giant Rat	[86]
P. ursinus (A)	Sulawesi Bear Rat	
Genus *Paulamys*		
P. naso (A)	Flores Long-nosed Rat	[82]
Genus *Pelomys*	Groove-toothed Swamp Rats	[16]
P. campanae (E)	Bell Groove-toothed Swamp Rat	
P. fallax (E)	Creek Groove-toothed Swamp Rat	
P. hopkinsi (E)	Hopkins's Groove-toothed Swamp Rat	
P. isseli (E)	Issel's Groove-toothed Swamp Rat	
P. minor (E)	Least Groove-toothed Swamp Rat	
Genus *Phloeomys*	Giant Cloud Rats	
P. cumingi (Or)	Southern Luzon Giant Cloud Rat	[48]

P. pallidus (Or)	Northern Luzon Giant Cloud Rat	[48]
Genus *Pithecheir*	Sunda Tree Rats	
P. melanurus (Or)	Red Tree Rat	[109]
P. parvus (Or)	Malayan Tree Rat	[16]
Genus *Pogonomelomys*	New Guinean Brush Mice	
P. bruijni (A)	Lowland Brush Mouse	[16]
P. mayeri (A)	Shaw Mayer's Brush Mouse	[16]
P. sevia (A)	Highland Brush Mouse	[16]
Genus *Pogonomys*	Prehensile-tailed Tree Mice	
P. championi (A)	Champion's Tree Mouse	[28]
P. loriae (A)	Large Tree Mouse	[28]
P. macrourus (A)	Chestnut Tree Mouse	[28]
P. sylvestris (A)	Gray-bellied Tree Mouse	[28]
Genus *Praomys*	African Soft-furred Mice	
P. delectorum (E)	Delectable Soft-furred Mouse	
P. hartwigi (E)	Hartweg's Soft-furred Mouse	
P. jacksoni (E)	Jackson's Soft-furred Mouse	[41]
P. minor (E)	Least Soft-furred Mouse	
P. misonnei (E)	Misonne's Soft-furred Mouse	
P. morio (E)	Cameroon Soft-furred Mouse	
P. mutoni (E)	Muton's Soft-furred Mouse	
P. rostratus (E)	Forest Soft-furred Mouse	
P. tullbergi (E)	Tullberg's Soft-furred Mouse	
Genus *Pseudohydromys*	New Guinean Shrew Mice	
P. murinus (A)	Eastern Shrew Mouse	[28]
P. occidentalis (A)	Western Shrew Mouse	[28]
Genus *Pseudomys*	Australian Mice	[16]
P. albocinereus (A)	Ash-gray Mouse	[110]
P. apodemoides (A)	Silky Mouse	[110]
P. australis (A)	Plains Mouse	[110]
P. bolami (A)	Bolam's Mouse	[110]
P. chapmani (A)	Western Pebble-mound Mouse	[110]
P. delicatulus (A)	Little Native Mouse	[16]

P. desertor (A)	Brown Desert Mouse	[16]
P. fieldi (A)	Alice Springs Mouse	[16]
P. fumeus (A)	Smoky Mouse	[110]
P. fuscus (A)	Broad-toothed Mouse	
P. glaucus (A)	Blue-gray Mouse	[109]
P. gouldii (A)	Gould's Mouse	[16]
P. gracilicaudatus (A)	Eastern Chestnut Mouse	[16]
P. hermannsburgensis (A)	Sandy Inland Mouse	[16]
P. higginsi (A)	Long-tailed Mouse	[110]
P. johnsoni (A)	Central Pebble-mound Mouse	[110]
P. laborifex (A)	Kimberly Mouse	[110]
P. nanus (A)	Western Chestnut Mouse	[110]
P. novaehollandiae (A)	New Holland Mouse	[110]
P. occidentalis (A)	Western Mouse	[110]
P. oralis (A)	Hastings River Mouse	[110]
P. patrius (A)	Country Mouse	
P. pilligaensis (A)	Pilliga Mouse	[110]
P. praeconis (A)	Shark Bay Mouse	[16]
P. shortridgei (A)	Heath Rat	[110]
Genus *Rattus*	Old World Rats	[16]
R. adustus (Or)	Sunburned Rat	
R. annandalei (Or)	Annandale's Rat	[16]
R. argentiventer (Or, A)	Rice-field Rat	[16]
R. baluensis (Or)	Summit Rat	[16]
R. bontanus (A)	Bonthain Rat	
R. burrus (Or)	Nonsense Rat	
R. colletti (A)	Dusky Rat	[110]
R. elaphinus (A)	Sula Rat	[29]
R. enganus (Or)	Enggano Rat	
R. everetti (Or)	Philippine Forest Rat	
R. exulans (Oc, Or, A)	Polynesian Rat	[16]
R. feliceus (A)	Spiny Ceram Rat	[29]
R. foramineus (A)	Hole Rat	
R. fuscipes (A)	Bush Rat	[16]
R. giluwensis (A)	Giluwe Rat	[28]
R. hainaldi (A)	Hainald's Rat	
R. hoffmanni (A)	Hoffmann's Rat	
R. hoogerwerfi (Or)	Hoogerwerf's Rat	

R. jobiensis (A)	Japen Rat	[29]
R. koopmani (A)	Koopman's Rat	
R. korinchi (Or)	Korinch's Rat	
R. leucopus (A)	Cape York Rat	[110]
R. losea (P, Or)	Lesser Rice-field Rat	[16]
R. lugens (Or)	Mentawai Rat	
R. lutreolus (A)	Australian Swamp Rat	[16]
R. macleari (Oc)	MacLear's Rat	[37]
R. marmosurus (A)	Opossum Rat	
R. mindorensis (Or)	Mindoro Black Rat	
R. mollicomulus (A)	Little Soft-furred Rat	
R. montanus (Or)	Nillu Rat	[93]
R. mordax (A)	Eastern Rat	[28]
R. morotaiensis (A)	Molaccan Prehensile-tailed Rat	[29]
R. nativitatis (Oc)	Bulldog Rat	[37]
R. nitidus (P, Or)	Himalayan Field Rat	[48]
R. norvegicus (All)	Brown Rat	[16]
R. novaeguineae (A)	New Guinean Rat	[28]
R. osgoodi (Or)	Osgood's Rat	
R. palmarum (Or)	Palm Rat	
R. pelurus (A)	Peleng Rat	
R. praetor (A)	Spiny Rat	[29]
R. ranjiniae (Or)	Kerala Rat	
R. rattus (All)	House Rat	[16]
R. sanila (A)	New Ireland Rat	
R. sikkimensis (P, Or)	Sikkim Rat	
R. simalurensis (Or)	Simalur Rat	
R. sordidus (A)	Dusky Field Rat	[117]
R. steini (A)	Stein's Rat	
R. stoicus (Or)	Andaman Rat	
R. tanezumi (P, Or)	Tanezumi Rat	
R. tawitawiensis (Or)	Tawi-tawi Forest Rat	[48]
R. timorensis (A)	Timor Rat	
R. tiomanicus (Or)	Malayan Field Rat	[16]
R. tunneyi (A)	Pale Field Rat	[110]
R. turkestanicus (P, Or)	Turkestan Rat	[16]
R. villosissimus (A)	Long-haired Rat	[110]
R. xanthurus (A)	Yellow-tailed Rat	

Genus *Rhabdomys*
R. *pumilio* (E)   Four-striped Grass Mouse   [16]
Genus *Rhynchomys*   Shrew Rats
R. *isarogensis* (Or)   Isarog Shrew Rat   [48]
R. *soricoides* (Or)   Mt. Data Shrew Rat   [16]
Genus *Solomys*   Naked-tailed Rats   [16]
S. *ponceleti* (A)   Poncelet's Naked-tailed
                          Rat
S. *salamonis* (A)   Florida Naked-tailed Rat
S. *salebrosus* (A)   Bougainville Naked-tailed
                          Rat
S. *sapientis* (A)   Isabel Naked-tailed Rat
S. *spriggsarum* (A)   Buka Naked-tailed Rat
Genus *Spelaeomys*
S. *florensis* (A)   Flores Cave Rat
Genus *Srilankamys*
S. *ohiensis* (Or)   Ohiya Rat   [16]
Genus *Stenocephalemys*   Ethiopian Narrow-headed
                          Rats   [86]
S. *albocaudata* (E)   Ethiopian Narrow-headed
                          Rat   [109]
S. *griseicauda* (E)   Gray-tailed Narrow-headed
                          Rat
Genus *Stenomys*   Slender Rats
S. *ceramicus* (A)   Ceram Rat   [16]
S. *niobe* (A)   Moss-forest Rat   [28]
S. *richardsoni* (A)   Glacier Rat   [28]
S. *vandeuseni* (A)   Van Deusen's Rat   [28]
S. *verecundus* (A)   Slender Rat   [28]
Genus *Stochomys*
S. *longicaudatus* (E)   Target Rat   [109]
Genus *Sundamys*   Giant Sunda Rats   [86]
S. *infraluteus* (Or)   Mountain Giant Rat   [109]
S. *maxi* (Or)   Bartels's Rat   [20]
S. *muelleri* (Or)   Müller's Giant Sunda Rat
Genus *Taeromys*   Sulawesi Rats
T. *arcuatus* (A)   Salokko Rat
T. *callitrichus* (A)   Lovely-haired Rat
T. *celebensis* (A)   Celebes Rat
T. *hamatus* (A)   Sulawesi Montane Rat

T. punicans (A)	Sulawesi Forest Rat	
T. taerae (A)	Tondano Rat	
Genus *Tarsomys*	Long-footed Rats	
T. apoensis (Or)	Long-footed Rat	[20]
T. echinatus (Or)	Spiny Long-footed Rat	
Genus *Tateomys*	Greater Sulawesian Shrew Rats	[86]
T. macrocercus (A)	Long-tailed Shrew Rat	
T. rhinogradoides (A)	Tate's Shrew Rat	
Genus *Thallomys*	Acacia Rats	
T. loringi (E)	Loring's Rat	
T. nigricauda (E)	Black-tailed Tree Rat	[107]
T. paedulcus (E)	Acacia Rat	[16]
T. shortridgei (E)	Shortridge's Rat	
Genus *Thamnomys*	Thicket Rats	[86]
T. kempi (E)	Kemp's Thicket Rat	
T. venustus (E)	Charming Thicket Rat	
Genus *Tokudaia*	Ryukyu Spiny Rats	[109]
T. osimensis (P)	Ryukyu Spiny Rat	[16]
T. muenninki (P)	Muennink's Spiny Rat	
Genus *Tryphomys*		
T. adustus (Or)	Luzon Short-nosed Rat	[48]
Genus *Uranomys*		
U. ruddi (E)	Rudd's Mouse	[16]
Genus *Uromys*	Giant Naked-tailed Rats	[86]
U. anak (A)	Giant Naked-tailed Rat	[109]
U. caudimaculatus (A)	Giant White-tailed Rat	[16]
U. hadrourus (A)	Masked White-tailed Rat	[110]
U. imperator (A)	Emperor Rat	[29]
U. neobritanicus (A)	Bismarck Giant Rat	[29]
U. porculus (A)	Guadalcanal Rat	[29]
U. rex (A)	King Rat	[29]
Genus *Vandeleuria*	Long-tailed Climbing Mice	[86]
V. nolthenii (Or)	Nolthenius's Long-tailed Climbing Mouse	
V. oleracea (P, Or)	Asiatic Long-tailed Climbing Mouse	
Genus *Vernaya*		
V. fulva (P, Or)	Red Climbing Mouse	

Genus *Xenuromys*		
X. *barbatus* (A)	Rock-dwelling Giant Rat	[28]
Genus *Xeromys*		
X. *myoides* (A)	False Water Rat	[109]
Genus *Zelotomys*	Broad-headed Mice	[86]
Z. *hildegardeae* (E)	Hildegarde's Broad-headed Mouse	
Z. *woosnami* (E)	Woosnam's Broad-headed Mouse	
Genus *Zyzomys*	Australian Rock Rats	[16]
Z. *argurus* (A)	Silver-tailed Rock Rat	
Z. *maini* (A)	Arnhem Land Rock Rat	[110]
Z. *palatilis* (A)	Carpentarian Rock Rat	[110]
Z. *pedunculatus* (A)	Central Rock Rat	[110]
Z. *woodwardi* (A)	Kimberly Rock Rat	[110]
Subfamily Myospalacinae		
Genus *Myospalax*	Zokors	[86]
M. *aspalax* (P)	False Zokor	
M. *epsilanus* (P)	Manchurian Zokor	
M. *fontanierii* (P)	Chinese Zokor	
M. *myospalax* (P)	Siberian Zokor	[16]
M. *psilurus* (P)	Transbaikal Zokor	
M. *rothschildi* (P)	Rothschild's Zokor	[16]
M. *smithii* (P)	Smith's Zokor	[16]
Subfamily Mystromyinae		
Genus *Mystromys*		
M. *albicaudatus* (E)	White-tailed Mouse	[107]
Subfamily Nesomyinae		
Genus *Brachytarsomys*		
B. *albicauda* (E)	White-tailed Rat	
Genus *Brachyuromys*	Short-tailed Rats	
B. *betsileoensis* (E)	Betsileo Short-tailed Rat	
B. *ramirohitra* (E)	Gregarious Short-tailed Rat	
Genus *Eliurus*	Tufted-tailed Rats	
E. *majori* (E)	Major's Tufted-tailed Rat	
E. *minor* (E)	Lesser Tufted-tailed Rat	
E. *myoxinus* (E)	Dormouse Tufted-tailed Rat	
E. *penicillatus* (E)	White-tipped Tufted-tailed Rat	

E. tanala (E)	Tanala Tufted-tailed Rat	
E. webbi (E)	Webb's Tufted-tailed Rat	
Genus *Gymnuromys*		
G. roberti (E)	Voalavoanala	[86]
Genus *Hypogeomys*		
H. antimena (E)	Malagasy Giant Rat	[86]
Genus *Macrotarsomys*	Big-footed Mice	
M. bastardi (E)	Bastard Big-footed Mouse	
M. ingens (E)	Greater Big-footed Mouse	
Genus *Nesomys*		
N. rufus (E)	Island Mouse	
Subfamily Otomyinae		
Genus *Otomys*	Vlei Rats	[109]
O. anchietae (E)	Angolan Vlei Rat	
O. angoniensis (E)	Angoni Vlei Rat	[16]
O. denti (E)	Dent's Vlei Rat	
O. irroratus (E)	Vlei Rat	[16]
O. laminatus (E)	Laminate Vlei Rat	[16]
O. maximus (E)	Large Vlei Rat	[16]
O. occidentalis (E)	Western Vlei Rat	
O. saundersiae (E)	Saunders's Vlei Rat	[16]
O. sloggetti (E)	Sloggett's Vlei Rat	
O. tropicalis (E)	Tropical Vlei Rat	
O. typus (E)	Typical Vlei Rat	
O. unisulcatus (E)	Bush Vlei Rat	
Genus *Parotomys*	Whistling Rats	
P. brantsii (E)	Brants's Whistling Rat	[16]
P. littledalei (E)	Littledale's Whistling Rat	[16]
Subfamily Petromyscinae		
Genus *Delanymys*		
D. brooksi (E)	Delany's Swamp Mouse	[109]
Genus *Petromyscus*	Rock Mice	[16]
P. barbouri (E)	Barbour's Rock Mouse	[107]
P. collinus (E)	Pygmy Rock Mouse	[16]
P. monticularis (E)	Brukkaros Pygmy Rock Mouse	[16]
P. shortridgei (E)	Shortridge's Rock Mouse	[107]
Subfamily Platacanthomyinae		
Genus *Platacanthomys*		
P. lasiurus (Or)	Malabar Spiny Dormouse	[16]

Genus *Typhlomys*	Chinese Pygmy Dormice	[109]
T. chapensis (Or)	Chapa Pygmy Dormouse	
T. cinereus (P)	Chinese Pygmy Dormouse	[16]
Subfamily Rhizomyinae		
Genus *Cannomys*		
C. badius (P, Or)	Lesser Bamboo Rat	[16]
Genus *Rhizomys*	Bamboo Rats	[16]
R. pruinosus (P, Or)	Hoary Bamboo Rat	[16]
R. sinensis (P, Or)	Chinese Bamboo Rat	[16]
R. sumatrensis (P, Or)	Large Bamboo Rat	[16]
Genus *Tachyoryctes*	African Mole Rats	[86]
T. ankoliae (E)	Ankole Mole Rat	
T. annectens (E)	Mianzini Mole Rat	
T. audax (E)	Audacious Mole Rat	
T. daemon (E)	Demon Mole Rat	
T. macrocephalus (E)	Big-headed Mole Rat	
T. naivashae (E)	Naivasha Mole Rat	
T. rex (E)	King Mole Rat	
T. ruandae (E)	Ruanda Mole Rat	
T. ruddi (E)	Rudd's Mole Rat	
T. spalacinus (E)	Embi Mole Rat	
T. splendens (E)	East African Mole Rat	[16]
Subfamily Sigmodontinae		
Genus *Abrawayaomys*		
A. ruschii (Neo)	Ruschi's Rat	
Genus *Aepeomys*	Montane Mice	
A. fuscatus (Neo)	Dusky Montane Mouse	
A. lugens (Neo)	Olive Montane Mouse	
Genus *Akodon*	Grass Mice	[86]
A. aerosus (Neo)	Highland Grass Mouse	
A. affinis (Neo)	Colombian Grass Mouse	
A. albiventer (Neo)	White-bellied Grass Mouse	
A. azarae (Neo)	Azara's Grass Mouse	
A. bogotensis (Neo)	Bogota Grass Mouse	
A. boliviensis (Neo)	Bolivian Grass Mouse	
A. budini (Neo)	Budin's Grass Mouse	
A. cursor (Neo)	Cursor Grass Mouse	
A. dayi (Neo)	Day's Grass Mouse	
A. dolores (Neo)	Dolorous Grass Mouse	

A. *fumeus* (Neo)	Smoky Grass Mouse	
A. *hershkovitzi* (Neo)	Hershkovitz's Grass Mouse	
A. *illuteus* (Neo)	Gray Grass Mouse	[7]
A. *iniscatus* (Neo)	Intelligent Grass Mouse	
A. *juninensis* (Neo)	Junin Grass Mouse	
A. *kempi* (Neo)	Kemp's Grass Mouse	
A. *kofordi* (Neo)	Koford's Grass Mouse	
A. *lanosus* (Neo)	Woolly Grass Mouse	
A. *latebricola* (Neo)	Ecuadorean Grass Mouse	
A. *lindberghi* (Neo)	Lindbergh's Grass Mouse	
A. *longipilis* (Neo)	Long-haired Grass Mouse	
A. *mansoensis* (Neo)	Manso Grass Mouse	
A. *markhami* (Neo)	Markham's Grass Mouse	
A. *mimus* (Neo)	Thespian Grass Mouse	
A. *molinae* (Neo)	Molina's Grass Mouse	
A. *mollis* (Neo)	Soft Grass Mouse	
A. *neocenus* (Neo)	Neuquen Grass Mouse	
A. *nigrita* (Neo)	Blackish Grass Mouse	
A. *olivaceus* (Neo)	Olive Grass Mouse	
A. *orophilus* (Neo)	El Dorado Grass Mouse	
A. *puer* (Neo)	Altiplano Grass Mouse	
A. *sanborni* (Neo)	Sanborn's Grass Mouse	
A. *sanctipaulensis* (Neo)	Sao Paulo Grass Mouse	
A. *serrensis* (Neo)	Serrado Mar Grass Mouse	
A. *siberiae* (Neo)	Cochabamba Grass Mouse	
A. *simulator* (Neo)	Gray-bellied Grass Mouse	[7]
A. *spegazzinii* (Neo)	Spegazzini's Grass Mouse	
A. *subfuscus* (Neo)	Puno Grass Mouse	
A. *surdus* (Neo)	Silent Grass Mouse	
A. *sylvanus* (Neo)	Forest Grass Mouse	
A. *toba* (Neo)	Chaco Grass Mouse	
A. *torques* (Neo)	Cloud Forest Grass Mouse	
A. *urichi* (Neo)	Northern Grass Mouse	
A. *varius* (Neo)	Variable Grass Mouse	
A. *xanthorhinus* (Neo)	Yellow-nosed Grass Mouse	
Genus *Andalgalomys*	Chaco Mice	
A. *olrogi* (Neo)	Olrog's Chaco Mouse	
A. *pearsoni* (Neo)	Pearson's Chaco Mouse	
Genus *Andinomys*		
A. *edax* (Neo)	Andean Mouse	[16]

Genus *Anotomys*
    A. leander (Neo)   Ecuador Fish-eating Rat   [16]
Genus *Auliscomys*   Big-eared Mice
    A. boliviensis (Neo)   Bolivian Big-eared Mouse
    A. micropus (Neo)   Southern Big-eared Mouse
    A. pictus (Neo)   Painted Big-eared Mouse
    A. sublimis (Neo)   Andean Big-eared Mouse
Genus *Baiomys*   American Pygmy Mice   [16]
    B. musculus (Nea, Neo)   Southern Pygmy Mouse   [98]
    B. taylori (Nea)   Northern Pygmy Mouse   [114]
Genus *Bibimys*   Crimson-nosed Rats   [86]
    B. chacoensis (Neo)   Chaco Crimson-nosed Rat
    B. labiosus (Neo)   Large-lipped Crimson-nosed Rat
    B. torresi (Neo)   Torres's Crimson-nosed Rat

Genus *Blarinomys*
    B. breviceps (Neo)   Brazilian Shrew Mouse   [26]
Genus *Bolomys*   Bolo Mice
    B. amoenus (Neo)   Pleasant Bolo Mouse
    B. lactens (Neo)   Rufous-bellied Bolo Mouse
    B. lasiurus (Neo)   Hairy-tailed Bolo Mouse
    B. obscurus (Neo)   Dark Bolo Mouse
    B. punctulatus (Neo)   Spotted Bolo Mouse
    B. temchuki (Neo)   Temchuk's Bolo Mouse
Genus *Calomys*   Vesper Mice   [16]
    C. boliviae (Neo)   Bolivian Vesper Mouse
    C. callidus (Neo)   Crafty Vesper Mouse
    C. callosus (Neo)   Large Vesper Mouse   [7]
    C. hummelincki (Neo)   Hummelinck's Vesper Mouse
    C. laucha (Neo)   Small Vesper Mouse
    C. lepidus (Neo)   Andean Vesper Mouse   [7]
    C. musculinus (Neo)   Drylands Vesper Mouse   [7]
    C. sorellus (Neo)   Peruvian Vesper Mouse
    C. tener (Neo)   Delicate Vesper Mouse
Genus *Chelemys*   Greater Long-clawed Mice   [86]
    C. macronyx (Neo)   Andean Long-clawed Mouse
    C. megalonyx (Neo)   Large Long-clawed Mouse

Genus *Chibchanomys*

    C. trichotis (Neo)   Chibchan Water Mouse   [86]

Genus *Chilomys*

    C. instans (Neo)   Colombian Forest Mouse   [16]

Genus *Chinchillula*

    C. sahamae (Neo)   Altiplano Chinchilla Mouse   [109]

Genus *Chroeomys*   Altiplano Mice

    C. andinus (Neo)   Andean Altiplano Mouse

    C. jelskii (Neo)   Jelski's Altiplano Mouse

Genus *Delomys*   Atlantic Forest Rats   [26]

    D. dorsalis (Neo)   Striped Atlantic Forest Rat

    D. sublineatus (Neo)   Pallid Atlantic Forest Rat

Genus *Eligmodontia*   Gerbil Mice

    E. moreni (Neo)   Monte Gerbil Mouse   [70]

    E. morgani (Neo)   Morgan's Gerbil Mouse

    E. puerulus (Neo)   Andean Gerbil Mouse   [70]

    E. typus (Neo)   Highland Gerbil Mouse

Genus *Euneomys*   Chinchilla Mice

    E. chinchilloides (Neo)   Patagonian Chinchilla
         Mouse   [109]

    E. fossor (Neo)   Burrowing Chinchilla
         Mouse

    E. mordax (Neo)   Biting Chinchilla Mouse

    E. petersoni (Neo)   Peterson's Chinchilla
         Mouse

Genus *Galenomys*

    G. garleppi (Neo)   Garlepp's Mouse

Genus *Geoxus*

    G. valdivianus (Neo)   Long-clawed Mole Mouse   [86]

Genus *Graomys*   Leaf-eared Mice

    G. domorum (Neo)   Pale Leaf-eared Mouse   [7]

    G. edithae (Neo)   Edith's Leaf-eared Mouse

    G. griseoflavus (Neo)   Gray Leaf-eared Mouse   [7]

Genus *Habromys*   Crested-tailed Deer Mice   [86]

    H. chinanteco (Nea)   Chinanteco Deer Mouse   [38]

    H. lepturus (Nea)   Slender-tailed Deer Mouse   [109]

    H. lophurus (Nea, Neo)   Crested-tailed Deer Mouse   [98]

    H. simulatus (Nea)   Jico Deer Mouse   [16]

Genus *Hodomys*

    H. alleni (Nea)   Allen's Woodrat   [16]

Genus *Holochilus*	Marsh Rats	[16]
H. brasiliensis (Neo)	Web-footed Marsh Rat	[109]
H. chacarius (Neo)	Chaco Marsh Rat	
H. magnus (Neo)	Greater Marsh Rat	
H. sciureus (Neo)	Marsh Rat	[24]
Genus *Ichthyomys*	Crab-eating Rats	[26]
I. hydrobates (Neo)	Crab-eating Rat	
I. pittieri (Neo)	Pittier's Crab-eating Rat	
I. stolzmanni (Neo)	Stolzmann's Crab-eating Rat	
I. tweedii (Neo)	Tweedy's Crab-eating Rat	
Genus *Irenomys*		
I. tarsalis (Neo)	Chilean Climbing Mouse	
Genus *Isthmomys*	Isthmus Rats	[26]
I. flavidus (Neo)	Yellow Isthmus Rat	
I. pirrensis (Neo)	Mt. Pirri Isthmus Rat	
Genus *Juscelinomys*	Juscelin's Mice	
J. candango (Neo)	Candango Mouse	
J. vulpinus (Neo)	Molelike Mouse	
Genus *Kunsia*	South American Giant Rats	[86]
K. fronto (Neo)	Fossorial Giant Rat	
K. tomentosus (Neo)	Woolly Giant Rat	
Genus *Lenoxus*		
L. apicalis (Neo)	Andean Rat	[109]
Genus *Megadontomys*	Giant Deer Mice	
M. cryophilus (Nea)	Oaxaca Giant Deer Mouse	
M. nelsoni (Nea)	Nelson's Giant Deer Mouse	
M. thomasi (Nea)	Thomas's Giant Deer Mouse	
Genus *Megalomys*	West Indian Giant Rice Rats	[86]
M. desmarestii (Neo)	Antillean Giant Rice Rat	[16]
M. luciae (Neo)	Santa Lucia Giant Rice Rat	[16]
Genus *Melanomys*	Dark Rice Rats	
M. caliginosus (Neo)	Dusky Rice Rat	[16]
M. robustulus (Neo)	Robust Dark Rice Rat	
M. zunigae (Neo)	Zuniga's Dark Rice Rat	
Genus *Microryzomys*	Small Rice Rats	
M. altissimus (Neo)	Highland Small Rice Rat	
M. minutus (Neo)	Forest Small Rice Rat	

Genus *Neacomys*	Bristly Mice	[38]
N. guianae (Neo)	Guiana Bristly Mouse	
N. pictus (Neo)	Painted Bristly Mouse	[38]
N. spinosus (Neo)	Bristly Mouse	[109]
N. tenuipes (Neo)	Narrow-footed Bristly Mouse	
Genus *Nectomys*	Neotropical Water Rats	[86]
N. palmipes (Neo)	Trinidad Water Rat	
N. parvipes (Neo)	Small-footed Water Rat	
N. squamipes (Neo)	South American Water Rat	[16]
Genus *Nelsonia*	Diminutive Woodrats	[86]
N. goldmani (Nea)	Nelson and Goldman's Woodrat	
N. neotomodon (Nea)	Diminutive Woodrat	[16]
Genus *Neotoma*	Woodrats	[16]
N. albigula (Nea)	White-throated Woodrat	[114]
N. angustapalata (Nea)	Tamaulipan Woodrat	[38]
N. anthonyi (Nea)	Anthony's Woodrat	[38]
N. bryanti (Nea)	Bryant's Woodrat	[38]
N. bunkeri (Nea)	Bunker's Woodrat	[38]
N. chrysomelas (Neo)	Nicaraguan Woodrat	[38]
N. cinerea (Nea)	Bushy-tailed Woodrat	[114]
N. devia (Nea)	Arizona Woodrat	[114]
N. floridana (Nea)	Eastern Woodrat	[114]
N. fuscipes (Nea)	Dusky-footed Woodrat	[114]
N. goldmani (Nea)	Goldman's Woodrat	[38]
N. lepida (Nea)	Desert Woodrat	[114]
N. martinensis (Nea)	San Martin Island Woodrat	[38]
N. mexicana (Neo, Nea)	Mexican Woodrat	[114]
N. micropus (Nea)	Southern Plains Woodrat	[114]
N. nelsoni (Nea)	Nelson's Woodrat	[38]
N. palatina (Nea)	Bolaños Woodrat	[38]
N. phenax (Nea)	Sonoran Woodrat	[38]
N. stephensi (Nea)	Stephens's Woodrat	[114]
N. varia (Nea)	Turner Island Woodrat	[38]
Genus *Neotomodon*		
N. alstoni (Nea)	Mexican Volcano Mouse	[86]
Genus *Neotomys*		
N. ebriosus (Neo)	Andean Swamp Rat	[16]

Genus *Nesoryzomys* — Galapagos Mice [16]
N. *darwini* (Neo) — Darwin's Galapagos Mouse
N. *fernandinae* (Neo) — Fernandina Galapagos Mouse
N. *indefessus* (Neo) — Indefatigable Galapagos Mouse
N. *swarthi* (Neo) — Santiago Galapagos Mouse
Genus *Neusticomys* — Fish-eating Rats
N. *monticolus* (Neo) — Montane Fish-eating Rat
N. *mussoi* (Neo) — Musso's Fish-eating Rat
N. *oyapocki* (Neo) — Oyapock's Fish-eating Rat
N. *peruviensis* (Neo) — Peruvian Fish-eating Rat [16]
N. *venezuelae* (Neo) — Venezuelan Fish-eating Rat [16]
Genus *Notiomys*
N. *edwardsii* (Neo) — Edwards's Long-clawed Mouse

Genus *Nyctomys*
N. *sumichrasti* (Nea, Neo) — Vesper Rat [26]
Genus *Ochrotomys*
O. *nuttalli* (Nea) — Golden Mouse [114]
Genus *Oecomys* — Arboreal Rice Rats [26]
O. *bicolor* (Neo) — Bicolored Arboreal Rice Rat
O. *cleberi* (Neo) — Cleber's Arboreal Rice Rat
O. *concolor* (Neo) — Unicolored Arboreal Rice Rat
O. *flavicans* (Neo) — Yellow Arboreal Rice Rat
O. *mamorae* (Neo) — Mamore Arboreal Rice Rat
O. *paricola* (Neo) — Brazilian Arboreal Rice Rat
O. *phaeotis* (Neo) — Dusky Arboreal Rice Rat
O. *rex* (Neo) — King Arboreal Rice Rat
O. *roberti* (Neo) — Robert's Arboreal Rice Rat
O. *rutilus* (Neo) — Red Arboreal Rice Rat
O. *speciosus* (Neo) — Arboreal Rice Rat
O. *superans* (Neo) — Foothill Arboreal Rice Rat
O. *trinitatis* (Neo) — Trinidad Arboreal Rice Rat
Genus *Oligoryzomys* — Pygmy Rice Rats [26]
O. *andinus* (Neo) — Andean Pygmy Rice Rat
O. *arenalis* (Neo) — Sandy Pygmy Rice Rat
O. *chacoensis* (Neo) — Chacoan Pygmy Rice Rat
O. *delticola* (Neo) — Delta Pygmy Rice Rat

O. destructor (Neo)	Destructive Pygmy Rice Rat	
O. eliurus (Neo)	Brazilian Pygmy Rice Rat	
O. flavescens (Neo)	Yellow Pygmy Rice Rat	
O. fulvescens (Nea, Neo)	Fulvous Pygmy Rice Rat	
O. griseolus (Neo)	Grayish Pygmy Rice Rat	
O. longicaudatus (Neo)	Long-tailed Pygmy Rice Rat	
O. magellanicus (Neo)	Magellanic Pygmy Rice Rat	
O. microtis (Neo)	Small-eared Pygmy Rice Rat	
O. nigripes (Neo)	Black-footed Pygmy Rice Rat	
O. vegetus (Neo)	Sprightly Pygmy Rice Rat	
O. victus (Neo)	St. Vincent Pygmy Rice Rat	
Genus *Onychomys*	Grasshopper Mice	[16]
O. arenicola (Nea)	Mearns's Grasshopper Mouse	[114]
O. leucogaster (Nea)	Northern Grasshopper Mouse	[114]
O. torridus (Nea)	Southern Grasshopper Mouse	[114]
Genus *Oryzomys*	Rice Rats	[16]
O. albigularis (Neo)	Tomes's Rice Rat	[16]
O. alfaroi (Nea, Neo)	Alfaro's Rice Rat	[38]
O. auriventer (Neo)	Ecuadorean Rice Rat	
O. balneator (Neo)	Peruvian Rice Rat	
O. bolivaris (Neo)	Bolivar Rice Rat	
O. buccinatus (Neo)	Paraguayan Rice Rat	
O. capito (Neo)	Large-headed Rice Rat	
O. chapmani (Nea)	Chapman's Rice Rat	[31]
O. couesi (Nea, Neo)	Coues's Rice Rat	[114]
O. devius (Neo)	Boquete Rice Rat	[38]
O. dimidiatus (Neo)	Thomas's Rice Rat	
O. galapagoensis (Neo)	Galapagos Rice Rat	
O. gorgasi (Neo)	Gorgas's Rice Rat	
O. hammondi (Neo)	Hammond's Rice Rat	
O. intectus (Neo)	Colombian Rice Rat	
O. intermedius (Neo)	Intermediate Rice Rat	
O. keaysi (Neo)	Keays's Rice Rat	
O. kelloggi (Neo)	Kellogg's Rice Rat	
O. lamia (Neo)	Monster Rice Rat	
O. legatus (Neo)	Big-headed Rice Rat	[70]

O. levipes (Neo)	Light-footed Rice Rat	
O. macconnelli (Neo)	MacConnell's Rice Rat	
O. melanotis (Nea)	Black-eared Rice Rat	[38]
O. nelsoni (Nea)	Nelson's Rice Rat	[38]
O. nitidus (Neo)	Elegant Rice Rat	
O. oniscus (Neo)	Sowbug Rice Rat	
O. palustris (Nea)	Marsh Rice Rat	[114]
O. polius (Neo)	Gray Rice Rat	
O. ratticeps (Neo)	Rat-headed Rice Rat	
O. rhabdops (Nea, Neo)	Striped Rice Rat	
O. rostratus (Nea, Neo)	Long-nosed Rice Rat	
O. saturatior (Nea, Neo)	Cloud Forest Rice Rat	
O. subflavus (Neo)	Terraced Rice Rat	
O. talamancae (Neo)	Talamancan Rice Rat	[38]
O. xantheolus (Neo)	Yellowish Rice Rat	
O. yunganus (Neo)	Yungas Rice Rat	
Genus Osgoodomys		
O. banderanus (Nea)	Michoacan Deer Mouse	[38]
Genus Otonyctomys		
O. hatti (Nea, Neo)	Hatt's Vesper Rat	
Genus Ototylomys		
O. phyllotis (Nea, Neo)	Big-eared Climbing Rat	[26]
Genus Oxymycterus	Hocicudos	[50]
O. akodontius (Neo)	Argentine Hocicudo	
O. angularis (Neo)	Angular Hocicudo	
O. delator (Neo)	Spy Hocicudo	
O. hiska (Neo)	Small Hocicudo	
O. hispidus (Neo)	Hispid Hocicudo	
O. hucucha (Neo)	Quechuan Hocicudo	
O. iheringi (Neo)	Ihering's Hocicudo	
O. inca (Neo)	Incan Hocicudo	
O. nasutus (Neo)	Long-nosed Hocicudo	
O. paramensis (Neo)	Paramo Hocicudo	
O. roberti (Neo)	Robert's Hocicudo	
O. rufus (Neo)	Red Hocicudo	
Genus Peromyscus	Deer Mice	[16]
P. attwateri (Nea)	Texas Mouse	[114]
P. aztecus (Nea, Neo)	Aztec Mouse	[98]
P. boylii (Nea)	Brush Mouse	[114]
P. bullatus (Nea)	Perote Mouse	[38]

P. californicus (Nea)	California Mouse	[114]
P. caniceps (Nea)	Burt's Deer Mouse	[38]
P. crinitus (Nea)	Canyon Mouse	[114]
P. dickeyi (Nea)	Dickey's Deer Mouse	[38]
P. difficilis (Nea)	Zacatecan Deer Mouse	[38]
P. eremicus (Nea)	Cactus Mouse	[114]
P. eva (Nea)	Eva's Desert Mouse	[38]
P. furvus (Nea)	Blackish Deer Mouse	[38]
P. gossypinus (Nea)	Cotton Mouse	[114]
P. grandis (Neo)	Big Deer Mouse	[38]
P. gratus (Nea)	Osgood's Mouse	[114]
P. guardia (Nea)	Angel Island Mouse	[38]
P. guatemalensis (Nea, Neo)	Guatemalan Deer Mouse	[38]
P. gymnotis (Nea, Neo)	Naked-eared Deer Mouse	[38]
P. hooperi (Nea)	Hooper's Mouse	[16]
P. interparietalis (Nea)	San Lorenzo Mouse	[16]
P. leucopus (Nea)	White-footed Mouse	[114]
P. levipes (Nea, Neo)	Nimble-footed Mouse	
P. madrensis (Nea)	Tres Marias Island Mouse	[16]
P. maniculatus (Nea)	Deer Mouse	[114]
P. mayensis (Neo)	Maya Mouse	[16]
P. megalops (Nea)	Brown Deer Mouse	[38]
P. mekisturus (Nea)	Puebla Deer Mouse	[38]
P. melanocarpus (Nea)	Zempoaltepec Deer Mouse	[38]
P. melanophrys (Nea)	Plateau Mouse	[38]
P. melanotis (Nea)	Black-eared Mouse	[54]
P. melanurus (Nea)	Black-tailed Mouse	[16]
P. merriami (Nea)	Mesquite Mouse	[52]
P. mexicanus (Nea, Neo)	Mexican Deer Mouse	[38]
P. nasutus (Nea)	Northern Rock Mouse	[114]
P. ochraventer (Nea)	El Carrizo Deer Mouse	[38]
P. oreas (Nea)	Columbian Mouse	[54]
P. pectoralis (Nea)	White-ankled Mouse	[114]
P. pembertoni (Nea)	Pemberton's Deer Mouse	[38]
P. perfulvus (Nea)	Marsh Mouse	[38]
P. polionotus (Nea)	Oldfield Mouse	[114]
P. polius (Nea)	Chihuahuan Mouse	[38]
P. pseudocrinitus (Nea)	False Canyon Mouse	[38]
P. sejugis (Nea)	Santa Cruz Mouse	[38]
P. simulus (Nea)	Nayarit Mouse	

P. sitkensis (Nea)	Sitka Mouse	[54]
P. slevini (Nea)	Slevin's Mouse	[38]
P. spicilegus (Nea)	Gleaning Mouse	[72]
P. stephani (Nea)	San Esteban Island Mouse	[38]
P. stirtoni (Neo)	Stirton's Deer Mouse	[38]
P. truei (Nea)	Pinyon Mouse	[114]
P. winkelmanni (Nea)	Winkelmann's Mouse	[16]
P. yucatanicus (Nea)	Yucatan Deer Mouse	[38]
P. zarhynchus (Nea)	Chiapan Deer Mouse	[38]
Genus *Phaenomys*		
P. ferrugineus (Neo)	Rio De Janeiro Arboreal Rat	[26]
Genus *Phyllotis*	Leaf-eared Mice	
P. amicus (Neo)	Friendly Leaf-eared Mouse	
P. andium (Neo)	Andean Leaf-eared Mouse	
P. bonaeriensis (Neo)	Buenos Aires Leaf-eared Mouse	
P. caprinus (Neo)	Capricorn Leaf-eared Mouse	
P. darwini (Neo)	Darwin's Leaf-eared Mouse	[70]
P. definitus (Neo)	Definitive Leaf-eared Mouse	
P. gerbillus (Neo)	Gerbil Leaf-eared Mouse	
P. haggardi (Neo)	Haggard's Leaf-eared Mouse	
P. magister (Neo)	Master Leaf-eared Mouse	
P. osgoodi (Neo)	Osgood's Leaf-eared Mouse	
P. osilae (Neo)	Bunchgrass Leaf-eared Mouse	[7]
P. wolffsohni (Neo)	Wolffsohn's Leaf-eared Mouse	
P. xanthopygus (Neo)	Yellow-rumped Leaf-eared Mouse	
Genus *Podomys*		
P. floridanus (Nea)	Florida Mouse	[114]
Genus *Podoxymys*		
P. roraimae (Neo)	Roraima Mouse	
Genus *Pseudoryzomys*		
P. simplex (Neo)	Brazilian False Rice Rat	
Genus *Punomys*		
P. lemminus (Neo)	Puna Mouse	[16]

Genus *Reithrodon*

R. *auritus* (Neo) | Bunny Rat | [7]

Genus *Reithrodontomys* | American Harvest Mice | [16]

R. *brevirostris* (Neo) | Short-nosed Harvest Mouse | [38]

R. *burti* (Nea) | Sonoran Harvest Mouse | [38]

R. *chrysopsis* (Nea) | Volcano Harvest Mouse | [38]

R. *creper* (Neo) | Chiriqui Harvest Mouse | [38]

R. *darienensis* (Neo) | Darien Harvest Mouse | [38]

R. *fulvescens* (Nea, Neo) | Fulvous Harvest Mouse | [114]

R. *gracilis* (Nea, Neo) | Slender Harvest Mouse | [38]

R. *hirsutus* (Nea) | Hairy Harvest Mouse | [38]

R. *humulis* (Nea) | Eastern Harvest Mouse | [114]

R. *megalotis* (Nea) | Western Harvest Mouse | [114]

R. *mexicanus* (Nea, Neo) | Mexican Harvest Mouse | [38]

R. *microdon* (Nea, Neo) | Small-toothed Harvest Mouse | [38]

R. *montanus* (Nea) | Plains Harvest Mouse | [114]

R. *paradoxus* (Neo) | Nicaraguan Harvest Mouse | [42]

R. *raviventris* (Nea) | Salt Marsh Harvest Mouse | [114]

R. *rodriguezi* (Neo) | Rodriguez's Harvest Mouse | [38]

R. *spectabilis* (Nea) | Cozumel Harvest Mouse | [38]

R. *sumichrasti* (Nea, Neo) | Sumichrast's Harvest Mouse | [38]

R. *tenuirostris* (Nea, Neo) | Narrow-nosed Harvest Mouse | [38]

R. *zacatecae* (Nea) | Zacatecas Harvest Mouse

Genus *Rhagomys*

R. *rufescens* (Neo) | Brazilian Arboreal Mouse | [26]

Genus *Rheomys* | Central American Water Mice | [26]

R. *mexicanus* (Nea) | Mexican Water Mouse | [16]

R. *raptor* (Neo) | Goldman's Water Mouse | [38]

R. *thomasi* (Nea, Neo) | Thomas's Water Mouse | [38]

R. *underwoodi* (Neo) | Underwood's Water Mouse | [38]

Genus *Rhipidomys* | American Climbing Mice | [16]

R. *austrinus* (Neo) | Southern Climbing Mouse

R. *caucensis* (Neo) | Cauca Climbing Mouse

R. *couesi* (Neo) | Coues's Climbing Mouse

R. *fulviventer* (Neo) | Buff-bellied Climbing Mouse

R. latimanus (Neo)	Broad-footed Climbing Mouse	
R. leucodactylus (Neo)	White-footed Climbing Mouse	[70]
R. macconnelli (Neo)	MacConnell's Climbing Mouse	
R. mastacalis (Neo)	Long-tailed Climbing Mouse	[109]
R. nitela (Neo)	Splendid Climbing Mouse	
R. ochrogaster (Neo)	Yellow-bellied Climbing Mouse	
R. scandens (Neo)	Mt. Pirri Climbing Mouse	[16]
R. venezuelae (Neo)	Venezuelan Climbing Mouse	
R. venustus (Neo)	Charming Climbing Mouse	
R. wetzeli (Neo)	Wetzel's Climbing Mouse	
Genus *Scapteromys*		
S. tumidus (Neo)	Swamp Rat	[50]
Genus *Scolomys*	Spiny Mice	[109]
S. melanops (Neo)	South American Spiny Mouse	[109]
S. ucayalensis (Neo)	Ucayali Spiny Mouse	
Genus *Scotinomys*	Brown Mice	[38]
S. teguina (Nea, Neo)	Alston's Brown Mouse	[38]
S. xerampelinus (Neo)	Chiriqui Brown Mouse	[38]
Genus *Sigmodon*	Cotton Rats	[16]
S. alleni (Nea)	Allen's Cotton Rat	[38]
S. alstoni (Neo)	Alston's Cotton Rat	
S. arizonae (Nea)	Arizona Cotton Rat	[114]
S. fulviventer (Nea)	Tawny-bellied Cotton Rat	[114]
S. hispidus (Nea, Neo)	Hispid Cotton Rat	[114]
S. inopinatus (Neo)	Unexpected Cotton Rat	
S. leucotis (Nea)	White-eared Cotton Rat	[38]
S. mascotensis (Nea)	Jaliscan Cotton Rat	[38]
S. ochrognathus (Nea)	Yellow-nosed Cotton Rat	[114]
S. peruanus (Neo)	Peruvian Cotton Rat	
Genus *Sigmodontomys*	Rice Water Rats	
S. alfari (Neo)	Alfaro's Rice Water Rat	
S. aphrastus (Neo)	Harris's Rice Water Rat	

Genus *Thalpomys* — Cerrado Mice
   T. cerradensis (Neo) — Cerrado Mouse
   T. lasiotis (Neo) — Hairy-eared Cerrado Mouse
Genus *Thomasomys* — Thomas's Oldfield Mice
   T. aureus (Neo) — Golden Oldfield Mouse
   T. baeops (Neo) — Beady-eyed Mouse
   T. bombycinus (Neo) — Silky Oldfield Mouse
   T. cinereiventer (Neo) — Ashy-bellied Oldfield Mouse
   T. cinereus (Neo) — Ash-colored Oldfield Mouse
   T. daphne (Neo) — Daphne's Oldfield Mouse
   T. eleusis (Neo) — Peruvian Oldfield Mouse
   T. gracilis (Neo) — Slender Oldfield Mouse
   T. hylophilus (Neo) — Woodland Oldfield Mouse
   T. incanus (Neo) — Inca Oldfield Mouse
   T. ischyurus (Neo) — Strong-tailed Oldfield Mouse
   T. kalinowskii (Neo) — Kalinowski's Oldfield Mouse
   T. ladewi (Neo) — Ladew's Oldfield Mouse
   T. laniger (Neo) — Butcher, Oldfield Mouse
   T. monochromos (Neo) — Unicolored Oldfield Mouse
   T. niveipes (Neo) — Snow-footed Oldfield Mouse
   T. notatus (Neo) — Distinguished Oldfield Mouse
   T. oreas (Neo) — Montane Oldfield Mouse
   T. paramorum (Neo) — Paramo Oldfield Mouse
   T. pyrrhonotus (Neo) — Thomas's Oldfield Mouse
   T. rhoadsi (Neo) — Rhoads's Oldfield Mouse
   T. rosalinda (Neo) — Rosalinda's Oldfield Mouse
   T. silvestris (Neo) — Forest Oldfield Mouse
   T. taczanowskii (Neo) — Taczanowski's Oldfield Mouse
   T. vestitus (Neo) — Dressy Oldfield Mouse
Genus *Tylomys* — Naked-tailed Climbing Rats [26]
   T. bullaris (Nea) — Chiapan Climbing Rat [38]
   T. fulviventer (Neo) — Fulvous-bellied Climbing Rat [38]
   T. mirae (Neo) — Mira Climbing Rat
   T. nudicaudus (Nea, Neo) — Peters's Climbing Rat [38]

T. panamensis (Neo)	Panamanian Climbing Rat	[38]
T. tumbalensis (Nea)	Tumbala Climbing Rat	[38]
T. watsoni (Neo)	Watson's Climbing Rat	[38]
Genus *Wiedomys*		
W. pyrrhorhinos (Neo)	Red-nosed Mouse	[16]
Genus *Wilfredomys*	Wilfred's Mice	
W. oenax (Neo)	Greater Wilfred's Mouse	
W. pictipes (Neo)	Lesser Wilfred's Mouse	
Genus *Xenomys*		
X. nelsoni (Nea)	Magdalena Rat	[38]
Genus *Zygodontomys*	Cane Mice	[16]
Z. brevicauda (Neo)	Short-tailed Cane Mouse	
Z. brunneus (Neo)	Brown Cane Mouse	
Subfamily Spalacinae		
Genus *Nannospalax*	Lesser Blind Mole Rats	
N. ehrenbergi (P)	Palestine Mole Rat	[109]
N. leucodon (P)	Lesser Mole Rat	[109]
N. nehringi (P)	Nehring's Blind Mole Rat	
Genus *Spalax*	Greater Blind Mole Rats	
S. arenarius (P)	Sandy Mole Rat	[111]
S. giganteus (P)	Giant Mole Rat	[111]
S. graecus (P)	Bukovin Mole Rat	[111]
S. microphthalmus (P)	Greater Mole Rat	[109]
S. zemni (P)	Podolsk Mole Rat	[111]
Family Anomaluridae	Scaly-tailed Squirrels	[16]
Subfamily Anomalurinae		
Genus *Anomalurus*	Scaly-tailed Flying Squirrels	[86]
A. beecrofti (E)	Beecroft's Scaly-tailed Squirrel	[109]
A. derbianus (E)	Lord Derby's Scaly-tailed Squirrel	
A. pelii (E)	Pel's Scaly-tailed Squirrel	[109]
A. pusillus (E)	Dwarf Scaly-tailed Squirrel	
Subfamily Zenkerellinae		
Genus *Idiurus*	Pygmy Scaly-tailed Flying Squirrels	[16]
I. macrotis (E)	Long-eared Scaly-tailed Flying Squirrel	
I. zenkeri (E)	Pygmy Scaly-tailed Flying Squirrel	

Genus *Zenkerella*		
Z. *insignis* (E)	Cameroon Scaly-tail	[63]
Family Pedetidae		
Genus *Pedetes*		
P. *capensis* (E)	Spring Hare	[16]
Family Ctenodactylidae	Gundis	[16]
Genus *Ctenodactylus*	Common Gundis	
C. *gundi* (P)	Gundi	[109]
C. *vali* (P)	Val's Gundi	
Genus *Felovia*		
F. *vae* (E)	Felou Gundi	[109]
Genus *Massoutiera*		
M. *mzabi* (P, E)	Mzab Gundi	[109]
Genus *Pectinator*		
P. *spekei* (E)	Speke's Pectinator	[109]
Family Myoxidae	Dormice	
Subfamily Graphiurinae		
Genus *Graphiurus*	African Dormice	[16]
G. *christyi* (E)	Christy's Dormouse	
G. *crassicaudatus* (E)	Jentink's Dormouse	[104]
G. *hueti* (E)	Huet's Dormouse	[109]
G. *kelleni* (E)	Kellen's Dormouse	[101]
G. *lorraineus* (E)	Lorrain Dormouse	
G. *microtis* (E)	Small-eared Dormouse	
G. *monardi* (E)	Monard's Dormouse	
G. *murinus* (E)	Woodland Dormouse	[16]
G. *ocularis* (E)	Spectacled Dormouse	[16]
G. *olga* (E)	Olga's Dormouse	
G. *parvus* (E)	Savanna Dormouse	
G. *platyops* (E)	Rock Dormouse	[16]
G. *rupicola* (E)	Stone Dormouse	
G. *surdus* (E)	Silent Dormouse	
Subfamily Leithiinae		
Genus *Dryomys*	Forest Dormice	[86]
D. *laniger* (P)	Woolly Dormouse	[16]
D. *nitedula* (P)	Forest Dormouse	[16]
D. *sichuanensis* (P)	Chinese Dormouse	[16]
Genus *Eliomys*	Garden Dormice	[86]
E. *melanurus* (P)	Asian Garden Dormouse	[43]
E. *quercinus* (P)	Garden Dormouse	[16]

Genus *Myomimus*	Mouse-tailed Dormice	[16]
M. personatus (P)	Masked Mouse-tailed Dormouse	
M. roachi (P)	Roach's Mouse-tailed Dormouse	
M. setzeri (P)	Setzer's Mouse-tailed Dormouse	
Genus *Selevinia*		
S. betpakdalaensis (P)	Desert Dormouse	[16]
Subfamily Myoxinae		
Genus *Glirulus*		
G. japonicus (P)	Japanese Dormouse	[16]
Genus *Muscardinus*		
M. avellanarius (P)	Hazel Dormouse	[16]
Genus *Myoxus*		
M. glis (P)	Fat Dormouse	[16]
Family Bathyergidae	Blesmols	
Genus *Bathyergus*	Dune Mole Rats	[16]
B. janetta (E)	Namaqua Dune Mole Rat	[16]
B. suillus (E)	Cape Dune Mole Rat	[16]
Genus *Cryptomys*	Common Mole Rats	[86]
C. bocagei (E)	Bocage's Mole Rat	[101]
C. damarensis (E)	Damara Mole Rat	[16]
C. foxi (E)	Nigerian Mole Rat	[104]
C. hottentotus (E)	African Mole Rat	[109]
C. mechowi (E)	Mechow's Mole Rat	
C. ochraceocinereus (E)	Ochre Mole Rat	[16]
C. zechi (E)	Togo Mole Rat	[104]
Genus *Georychus*		
G. capensis (E)	Cape Mole Rat	[16]
Genus *Heliophobius*		
H. argenteocinereus (E)	Silvery Mole Rat	[16]
Genus *Heterocephalus*		
H. glaber (E)	Naked Mole Rat	[16]
Family Hystricidae	Old World Porcupines	[16]
Genus *Atherurus*	Brush-tailed Porcupines	[16]
A. africanus (E)	African Brush-tailed Porcupine	[16]
A. macrourus (P, Or)	Asiatic Brush-tailed Porcupine	[16]

Genus *Hystrix*	Short-tailed Porcupines	[16]
H. africaeaustralis (E)	Cape Porcupine	[109]
H. brachyura (P, Or)	Malayan Porcupine	[109]
H. crassispinis (Or)	Thick-spined Porcupine	[91]
H. cristata (P, E)	Crested Porcupine	[16]
H. indica (P, Or)	Indian Crested Porcupine	[109]
H. javanica (Or, A)	Sunda Porcupine	[20]
H. pumila (Or)	Indonesian Porcupine	[109]
H. sumatrae (Or)	Sumatran Porcupine	
Genus *Trichys*		
T. fasciculata (Or)	Long-tailed Porcupine	[16]
Family Petromuridae		
Genus *Petromus*		
P. typicus (E)	Dassie Rat	[16]
Family Thryonomyidae	Cane Rats	[16]
Genus *Thryonomys*		
T. gregorianus (E)	Lesser Cane Rat	[16]
T. swinderianus (E)	Greater Cane Rat	[16]
Family Erethizontidae	New World Porcupines	[16]
Genus *Coendou*	Prehensile-tailed Porcupines	[86]
C. bicolor (Neo)	Bicolor-spined Porcupine	[26]
C. koopmani (Neo)	Koopman's Porcupine	
C. prehensilis (Neo)	Brazilian Porcupine	[26]
C. rothschildi (Neo)	Rothschild's Porcupine	[26]
Genus *Echinoprocta*		
E. rufescens (Neo)	Stump-tailed Porcupine	[26]
Genus *Erethizon*		
E. dorsatum (Nea)	North American Porcupine	[16]
Genus *Sphiggurus*	Hairy Dwarf Porcupines	
S. insidiosus (Neo)	Bahia Hairy Dwarf Porcupine	[26]
S. mexicanus (Nea, Neo)	Mexican Hairy Dwarf Porcupine	
S. pallidus (Neo)	Pallid Hairy Dwarf Porcupine	
S. spinosus (Neo)	Paraguay Hairy Dwarf Porcupine	[26]
S. vestitus (Neo)	Brown Hairy Dwarf Porcupine	[26]

S. villosus (Neo)	Orange-spined Hairy Dwarf Porcupine	[26]
Family Chinchillidae	Viscachas and Chinchillas	[16]
Genus *Chinchilla*	Chinchillas	[16]
C. brevicaudata (Neo)	Short-tailed Chinchilla	[109]
C. lanigera (Neo)	Chinchilla	[109]
Genus *Lagidium*	Mountain Viscachas	[86]
L. peruanum (Neo)	Northern Viscacha	
L. viscacia (Neo)	Southern Viscacha	
L. wolffsohni (Neo)	Wolffsohn's Viscacha	
Genus *Lagostomus*		
L. maximus (Neo)	Plains Viscacha	[16]
Family Dinomyidae		
Genus *Dinomys*		
D. branickii (Neo)	Pacarana	[26]
Family Caviidae	Guinea Pigs	[16]
Subfamily Caviinae		
Genus *Cavia*	Guinea Pigs	[16]
C. aperea (Neo)	Brazilian Guinea Pig	
C. fulgida (Neo)	Shiny Guinea Pig	
C. magna (Neo)	Greater Guinea Pig	
C. porcellus (Neo)	Guinea Pig	[109]
C. tschudii (Neo)	Montane Guinea Pig	
Genus *Galea*	Yellow-toothed Cavies	[86]
G. flavidens (Neo)	Yellow-toothed Cavy	
G. musteloides (Neo)	Common Yellow-toothed Cavy	[7]
G. spixii (Neo)	Spix's Yellow-toothed Cavy	
Genus *Kerodon*		
K. rupestris (Neo)	Rock Cavy	[16]
Genus *Microcavia*	Mountain Cavies	[86]
M. australis (Neo)	Southern Mountain Cavy	[109]
M. niata (Neo)	Andean Mountain Cavy	
M. shiptoni (Neo)	Shipton's Mountain Cavy	
Subfamily Dolichotinae		
Genus *Dolichotis*	Maras	[16]
D. patagonum (Neo)	Patagonian Mara	
D. salinicola (Neo)	Chacoan Mara	

Family Hydrochaeridae
  Genus *Hydrochaeris*
    H. hydrochaeris (Neo)

Family Dasyproctidae
  Genus *Dasyprocta*
    D. azarae (Neo)
    D. coibae (Neo)
    D. cristata (Neo)
    D. fuliginosa (Neo)
    D. guamara (Neo)
    D. kalinowskii (Neo)
    D. leporina (Neo)
    D. mexicana (Nea)
    D. prymnolopha (Neo)
    D. punctata (Nea, Neo)

    D. ruatanica (Neo)
  Genus *Myoprocta*
    M. acouchy (Neo)
    M. exilis (Neo)

Family Agoutidae
  Genus *Agouti*
    A. paca (Nea, Neo)
    A. taczanowskii (Neo)

Family Ctenomyidae
  Genus *Ctenomys*
    C. argentinus (Neo)
    C. australis (Neo)
    C. azarae (Neo)
    C. boliviensis (Neo)
    C. bonettoi (Neo)
    C. brasiliensis (Neo)
    C. colburni (Neo)
    C. conoveri (Neo)
    C. dorsalis (Neo)
    C. emilianus (Neo)
    C. frater (Neo)
    C. fulvus (Neo)
    C. haigi (Neo)

Capybara	[26]
Agoutis	[16]
Agoutis	[16]
Azara's Agouti	[26]
Coiban Agouti	
Crested Agouti	[109]
Black Agouti	[26]
Orinoco Agouti	
Kalinowski's Agouti	
Brazilian Agouti	[16]
Mexican Agouti	[16]
Black-rumped Agouti	[26]
Central American Agouti	[26]
Ruatan Island Agouti	[109]
Acouchis	[16]
Green Acouchi	[86]
Red Acouchi	[86]
Pacas	[86]
Paca	[109]
Mountain Paca	[109]
Tuco-tucos	[16]
Argentine Tuco-tuco	
Southern Tuco-tuco	
Azara's Tuco-tuco	
Bolivian Tuco-tuco	
Bonetto's Tuco-tuco	
Brazilian Tuco-tuco	
Colburn's Tuco-tuco	
Conover's Tuco-tuco	
Chacoan Tuco-tuco	
Emily's Tuco-tuco	
Forest Tuco-tuco	[70]
Tawny Tuco-tuco	
Haig's Tuco-tuco	

C. knighti (Neo)	Catamarca Tuco-tuco	[7]
C. latro (Neo)	Mottled Tuco-tuco	[7]
C. leucodon (Neo)	White-toothed Tuco-tuco	
C. lewisi (Neo)	Lewis's Tuco-tuco	
C. magellanicus (Neo)	Magellanic Tuco-tuco	
C. maulinus (Neo)	Maule Tuco-tuco	
C. mendocinus (Neo)	Mendoza Tuco-tuco	[70]
C. minutus (Neo)	Tiny Tuco-tuco	
C. nattereri (Neo)	Natterer's Tuco-tuco	
C. occultus (Neo)	Furtive Tuco-tuco	[7]
C. opimus (Neo)	Highland Tuco-tuco	[70]
C. pearsoni (Neo)	Pearson's Tuco-tuco	
C. perrensis (Neo)	Goya Tuco-tuco	
C. peruanus (Neo)	Peruvian Tuco-tuco	
C. pontifex (Neo)	San Luis Tuco-tuco	
C. porteousi (Neo)	Porteous's Tuco-tuco	
C. saltarius (Neo)	Salta Tuco-tuco	[70]
C. sericeus (Neo)	Silky Tuco-tuco	
C. sociabilis (Neo)	Social Tuco-tuco	
C. steinbachi (Neo)	Steinbach's Tuco-tuco	
C. talarum (Neo)	Talas Tuco-tuco	
C. torquatus (Neo)	Collared Tuco-tuco	
C. tuconax (Neo)	Robust Tuco-tuco	[7]
C. tucumanus (Neo)	Tucuman Tuco-tuco	
C. validus (Neo)	Strong Tuco-tuco	
Family Octodontidae	Octodonts	[86]
Genus *Aconaemys*	Rock Rats	[86]
A. fuscus (Neo)	Chilean Rock Rat	[16]
A. sagei (Neo)	Sage's Rock Rat	
Genus *Octodon*	Degus	[16]
O. bridgesi (Neo)	Bridges's Degu	
O. degus (Neo)	Degu	[109]
O. lunatus (Neo)	Moon-toothed Degu	
Genus *Octodontomys*		
O. gliroides (Neo)	Mountain Degu	[16]
Genus *Octomys*		
O. mimax (Neo)	Viscacha Rat	[16]
Genus *Spalacopus*		
S. cyanus (Neo)	Coruro	[16]

Genus *Tympanoctomys*		
T. barrerae (Neo)	Plains Viscacha Rat	
Family Abrocomidae	Chinchilla Rats	[16]
Genus *Abrocoma*		
A. bennetti (Neo)	Bennett's Chinchilla Rat	
A. boliviensis (Neo)	Bolivian Chinchilla Rat	
A. cinerea (Neo)	Ashy Chinchilla Rat	
Family Echimyidae	American Spiny Rats	[16]
Subfamily Chaetomyinae		
Genus *Chaetomys*		
C. subspinosus (Neo)	Bristle-spined Rat	
Subfamily Dactylomyinae		
Genus *Dactylomys*	Neotropical Bamboo Rats	
D. boliviensis (Neo)	Bolivian Bamboo Rat	
D. dactylinus (Neo)	Amazon Bamboo Rat	[26]
D. peruanus (Neo)	Peruvian Bamboo Rat	
Genus *Kannabateomys*		
K. amblyonyx (Neo)	Atlantic Bamboo Rat	
Genus *Olallamys*	Olalla Rats	
O. albicauda (Neo)	White-tailed Olalla Rat	
O. edax (Neo)	Greedy Olalla Rat	
Subfamily Echimyinae		
Genus *Diplomys*	Arboreal Soft-furred Spiny Rats	[109]
D. caniceps (Neo)	Arboreal Soft-furred Spiny Rat	[109]
D. labilis (Neo)	Rufous Tree Rat	[98]
D. rufodorsalis (Neo)	Red Crested Tree Rat	[26]
Genus *Echimys*	Spiny Tree Rats	[16]
E. blainvillei (Neo)	Golden Atlantic Tree Rat	[26]
E. braziliensis (Neo)	Red-nosed Tree Rat	[26]
E. chrysurus (Neo)	White-faced Tree Rat	[26]
E. dasythrix (Neo)	Drab Atlantic Tree Rat	[26]
E. grandis (Neo)	Giant Tree Rat	[26]
E. lamarum (Neo)	Pallid Atlantic Tree Rat	[26]
E. macrurus (Neo)	Long-tailed Tree Rat	
E. nigrispinus (Neo)	Black-spined Atlantic Tree Rat	[26]
E. pictus (Neo)	Painted Tree Rat	[26]

E. rhipidurus (Neo)	Peruvian Tree Rat	[26]
E. saturnus (Neo)	Dark Tree Rat	[26]
E. semivillosus (Neo)	Speckled Tree Rat	[26]
E. thomasi (Neo)	Giant Atlantic Tree Rat	[26]
E. unicolor (Neo)	Unicolored Tree Rat	
Genus *Isothrix*	Brush-tailed Rats	
I. bistriata (Neo)	Yellow-crowned Brush-tailed Rat	[26]
I. pagurus (Neo)	Plain Brush-tailed Rat	[26]
Genus *Makalata*		
M. armata (Neo)	Armored Spiny Rat	[109]
Subfamily Eumysopinae		
Genus *Carterodon*		
C. sulcidens (Neo)	Owl's Spiny Rat	[109]
Genus *Clyomys*	Lund's Spiny Rats	
C. bishopi (Neo)	Bishop's Fossorial Spiny Rat	
C. laticeps (Neo)	Broad-headed Spiny Rat	[109]
Genus *Euryzygomatomys*		
E. spinosus (Neo)	Guiara	[16]
Genus *Hoplomys*		
H. gymnurus (Neo)	Armored Rat	[26]
Genus *Lonchothrix*		
L. emiliae (Neo)	Tuft-tailed Spiny Tree Rat	[26]
Genus *Mesomys*	Spiny Tree Rats	
M. didelphoides (Neo)	Brazilian Spiny Tree Rat	
M. hispidus (Neo)	Spiny Tree Rat	[26]
M. leniceps (Neo)	Woolly-headed Spiny Tree Rat	
M. obscurus (Neo)	Dusky Spiny Tree Rat	
M. stimulax (Neo)	Surinam Spiny Tree Rat	
Genus *Proechimys*	Terrestrial Spiny Rats	[16]
P. albispinus (Neo)	White-spined Spiny Rat	
P. amphichoricus (Neo)	Venezuelan Spiny Rat	
P. bolivianus (Neo)	Bolivian Spiny Rat	
P. brevicauda (Neo)	Huallaga Spiny Rat	[90]
P. canicollis (Neo)	Colombian Spiny Rat	
P. cayennensis (Neo)	Cayenne Spiny Rat	[109]
P. chrysaeolus (Neo)	Boyaca Spiny Rat	

P. cuvieri (Neo)	Cuvier's Spiny Rat	
P. decumanus (Neo)	Pacific Spiny Rat	
P. dimidiatus (Neo)	Atlantic Spiny Rat	
P. goeldii (Neo)	Goeldi's Spiny Rat	
P. gorgonae (Neo)	Gorgona Spiny Rat	
P. guairae (Neo)	Guaira Spiny Rat	
P. gularis (Neo)	Ecuadoran Spiny Rat	
P. hendeei (Neo)	Hendee's Spiny Rat	
P. hoplomyoides (Neo)	Guyanan Spiny Rat	
P. iheringi (Neo)	Ihering's Spiny Rat	
P. longicaudatus (Neo)	Long-tailed Spiny Rat	
P. magdalenae (Neo)	Magdalena Spiny Rat	
P. mincae (Neo)	Minca Spiny Rat	
P. myosuros (Neo)	Mouse-tailed Spiny Rat	
P. oconnelli (Neo)	O'Connell's Spiny Rat	
P. oris (Neo)	Para Spiny Rat	
P. poliopus (Neo)	Gray-footed Spiny Rat	
P. quadruplicatus (Neo)	Napo Spiny Rat	
P. semispinosus (Neo)	Tome's Spiny Rat	[109]
P. setosus (Neo)	Hairy Spiny Rat	
P. simonsi (Neo)	Simon's Spiny Rat	
P. steerei (Neo)	Steere's Spiny Rat	
P. trinitatis (Neo)	Trinidad Spiny Rat	
P. urichi (Neo)	Sucre Spiny Rat	
P. warreni (Neo)	Warren's Spiny Rat	
Genus *Thrichomys*		
T. apereoides (Neo)	Punare	[109]
Subfamily Heteropsomyinae		
Genus *Boromys*	Cuban Cave Rats	
B. offella (Neo)	Oriente Cave Rat	
B. torrei (Neo)	Torre's Cave Rat	
Genus *Brotomys*	Edible Rats	
B. contractus (Neo)	Haitian Edible Rat	
B. voratus (Neo)	Hispaniolan Edible Rat	
Genus *Heteropsomys*	Hispaniolan Cave Rats	
H. antillensis (Neo)	Antillean Cave Rat	
H. insulans (Neo)	Insular Cave Rat	
Genus *Puertoricomys*		
P. corozalus (Neo)	Corozal Rat	

Family Capromyidae	Hutias	[86]
Subfamily Capromyinae		
Genus *Capromys*		
C. pilorides (Neo)	Desmarest's Hutia	[86]
Genus *Geocapromys*	Bahaman and Jamaican Hutias	[86]
G. brownii (Neo)	Brown's Hutia	[16]
G. ingrahami (Neo)	Bahamian Hutia	[86]
G. thoracatus (Neo)	Swan Island Hutia	[81]
Genus *Mesocapromys*	Sticknest Hutias	
M. angelcabrerai (Neo)	Cabrera's Hutia	
M. auritus (Neo)	Eared Hutia	
M. nanus (Neo)	Dwarf Hutia	[16]
M. sanfelipensis (Neo)	San Felipe Hutia	
Genus *Mysateles*	Long-tailed Cuban Hutias	[86]
M. garridoi (Neo)	Garrido's Hutia	
M. gundlachi (Neo)	Gundlach's Hutia	
M. melanurus (Neo)	Black-tailed Hutia	[109]
M. meridionalis (Neo)	Southern Hutia	
M. prehensilis (Neo)	Prehensile-tailed Hutia	[16]
Subfamily Hexolobodontinae		
Genus *Hexolobodon*		
H. phenax (Neo)	Imposter Hutia	
Subfamily Isolobodontinae		
Genus *Isolobodon*	Laminar-toothed Hutias	
I. montanus (Neo)	Montane Hutia	
I. portoricensis (Neo)	Puerto Rican Hutia	
Subfamily Plagiodontinae		
Genus *Plagiodontia*	Hispaniolan Hutias	[86]
P. aedium (Neo)	Hispaniolan Hutia	[16]
P. araeum (Neo)	San Rafael Hutia	
P. ipnaeum (Neo)	Samana Hutia	
Genus *Rhizoplagiodontia*		
R. lemkei (Neo)	Lemke's Hutia	
Family Heptaxodontidae	Key Mice	
Subfamily Clidomyinae		
Genus *Clidomys*	Key Mice	
C. osborni (Neo)	Osborn's Key Mouse	
C. parvus (Neo)	Small Key Mouse	

Subfamily Heptaxodontinae
  Genus *Amblyrhiza*
    A. inundata (Neo)   Blunt-toothed Mouse
  Genus *Elasmodontomys*
    E. obliquus (Neo)   Plate-toothed Mouse
  Genus *Quemisia*
    Q. gravis (Neo)   Twisted-toothed Mouse
Family Myocastoridae
  Genus *Myocastor*
    M. coypus (Neo)   Nutria   [109]

ORDER LAGOMORPHA   Rabbits, Hares, and
                    Pikas   [86]

Family Ochotonidae   Pikas   [16]
  Genus *Ochotona*   Pikas   [86]
    O. alpina (P)   Alpine Pika   [12]
    O. cansus (P)   Gansu Pika   [12]
    O. collaris (Nea)   Collared Pika   [12]
    O. curzoniae (P, Or)   Black-lipped Pika [12]
    O. dauurica (P)   Daurian Pika   [12]
    O. erythrotis (P)   Chinese Red Pika [12]
    O. forresti (P, Or)   Forrest's Pika   [12]
    O. gaoligongensis (P) Gaoligong Pika   [12]
    O. gloveri (P)   Glover's Pika   [12]
    O. himalayana (P, Or) Himalayan Pika   [12]
    O. hyperborea (P)   Northern Pika   [12]
    O. iliensis (P)   Ili Pika   [12]
    O. koslowi (P)   Kozlov's Pika   [12]
    O. ladacensis (P, Or)   Ladak Pika   [12]
    O. macrotis (P, Or)   Large-eared Pika [12]
    O. muliensis (P)   Muli Pika   [12]
    O. nubrica (P, Or)   Nubra Pika   [12]
    O. pallasi (P)   Pallas's Pika   [12]
    O. princeps (Nea)   American Pika   [12]
    O. pusilla (P)   Steppe Pika   [12]
    O. roylei (P, Or)   Royle's Pika   [12]
    O. rufescens (P)   Afghan Pika   [12]
    O. rutila (P)   Turkestan Red Pika [12]
    O. thibetana (P, Or)   Moupin Pika   [12]
    O. thomasi (P)   Thomas's Pika   [12]

Genus *Prolagus*
    P. sardus (P) — Sardinian Pika — [86]
Family Leporidae — Hares and Rabbits — [16]
  Genus *Brachylagus*
    B. idahoensis (Nea) — Pygmy Rabbit — [114]
  Genus *Bunolagus*
    B. monticularis (E) — Riverine Rabbit — [12]
  Genus *Caprolagus*
    C. hispidus (Or) — Hispid Hare — [12]
  Genus *Lepus* — Hares and Jackrabbits — [86]
    L. alleni (Nea) — Antelope Jackrabbit — [114]
    L. americanus (Nea) — Snowshoe Hare — [114]
    L. arcticus (Nea) — Arctic Hare — [114]
    L. brachyurus (P) — Japanese Hare — [12]
    L. californicus (Nea) — Black-tailed Jackrabbit — [114]
    L. callotis (Nea) — White-sided Jackrabbit — [114]
    L. capensis (P, E) — Cape Hare — [12]
    L. castroviejoi (P) — Broom Hare — [12]
    L. comus (P) — Yunnan Hare — [12]
    L. coreanus (P) — Korean Hare — [12]
    L. corsicanus (P) — Corsican Hare
    L. europaeus (P) — European Hare — [12]
    L. fagani (E) — Ethiopian Hare — [12]
    L. flavigularis (Nea) — Tehuantepec Jackrabbit — [12]
    L. granatensis (P) — Granada Hare
    L. hainanus (P) — Hainan Hare — [12]
    L. insularis (Nea) — Black Jackrabbit — [12]
    L. mandshuricus (P) — Manchurian Hare — [12]
    L. nigricollis (P, Or) — Indian Hare — [12]
    L. oiostolus (P, Or) — Woolly Hare — [12]
    L. othus (Nea, P) — Alaskan Hare — [114]
    L. pequensis (Or) — Burmese Hare — [12]
    L. saxatilis (E) — Scrub Hare — [12]
    L. sinensis (P, Or) — Chinese Hare — [12]
    L. starcki (E) — Ethiopian Highland Hare — [12]
    L. timidus (P) — Mountain Hare — [12]
    L. tolai (P) — Tolai Hare — [109]
    L. townsendii (Nea) — White-tailed Jackrabbit — [114]
    L. victoriae (E) — African Savanna Hare — [12]
    L. yarkandensis (P) — Yarkand Hare — [12]

Genus *Nesolagus*		
N. netscheri (Or)	Sumatran Rabbit	[12]
Genus *Oryctolagus*		
O. cuniculus (P)	European Rabbit	[12]
Genus *Pentalagus*		
P. furnessi (P)	Amami Rabbit	[12]
Genus *Poelagus*		
P. marjorita (E)	Bunyoro Rabbit	[12]
Genus *Pronolagus*	Red Rockhares	[109]
P. crassicaudatus (E)	Natal Red Rockhare	
P. randensis (E)	Jameson's Red Rockhare	[12]
P. rupestris (E)	Smith's Red Rockhare	[12]
Genus *Romerolagus*		
R. diazi (Nea)	Volcano Rabbit	[12]
Genus *Sylvilagus*	Cottontails	[16]
S. aquaticus (Nea)	Swamp Rabbit	[114]
S. audubonii (Nea)	Desert Cottontail	[114]
S. bachmani (Nea)	Brush Rabbit	[114]
S. brasiliensis (Nea, Neo)	Tapeti	
S. cunicularius (Nea)	Mexican Cottontail	[12]
S. dicei (Neo)	Dice's Cottontail	[12]
S. floridanus (Nea, Neo)	Eastern Cottontail	[114]
S. graysoni (Nea)	Tres Marias Cottontail	[12]
S. insonus (Nea)	Omilteme Cottontail	[12]
S. mansuetus (Nea)	San Jose Brush Rabbit	[12]
S. nuttallii (Nea)	Mountain Cottontail	[114]
S. palustris (Nea)	Marsh Rabbit	[114]
S. transitionalis (Nea)	New England Cottontail	[114]

ORDER MACROSCELIDEA	Elephant Shrews	[16]
Family Macroscelididae		
Genus *Elephantulus*	Long-eared Elephant Shrews	[86]
E. brachyrhynchus (E)	Short-snouted Elephant Shrew	[84]
E. edwardii (E)	Cape Elephant Shrew	[84]
E. fuscipes (E)	Dusky-footed Elephant Shrew	[84]
E. fuscus (E)	Dusky Elephant Shrew	[84]
E. intufi (E)	Bushveld Elephant Shrew	[84]

E. myurus (E)	Eastern Rock Elephant Shrew	[84]
E. revoili (E)	Somali Elephant Shrew	[84]
E. rozeti (P)	North African Elephant Shrew	[84]
E. rufescens (E)	Rufous Elephant Shrew	[84]
E. rupestris (E)	Western Rock Elephant Shrew	[84]
Genus *Macroscelides*		
M. proboscideus (E)	Short-eared Elephant Shrew	[84]
Genus *Petrodromus*		
P. tetradactylus (E)	Four-toed Elephant Shrew	[84]
Genus *Rhynchocyon*	Checkered Elephant Shrews	[86]
R. chrysopygus (E)	Golden-rumped Elephant Shrew	[84]
R. cirnei (E)	Checkered Elephant Shrew	[84]
R. petersi (E)	Black and Rufous Elephant Shrew	[84]

LITERATURE CITED

1   American Ornithologists' Union. 1983. *Check-list of North American birds*, 6th edition. Allen Press, Lawrence, Kansas.

2   Arroyo-Cabrales, J., and J. K. Jones, Jr. 1988. *Balantiopteryx io* and *Balantiopteryx infusca*. *Mammalian Species* 313:1–3.

3   Baillie, J., and B. Groombridge. 1996. *IUCN red list of threatened animals*. IUCN, Gland, Switzerland.

4   Banfield, A. W. F. 1974. *The mammals of Canada*. University of Toronto Press, Toronto, Canada.

5   Banks, R. C., R. W. McDiarmid, and A. L. Gardner. 1987. Checklist of Vertebrates of the United States, the U.S. Territories, and Canada. *United States Department of the Interior, Fish and Wildlife Service, Resource Publication* 166:1–79.

6   Barquez, R. M., N. P. Giannini, and M. A. Mares. 1993. *Guide to the bats of Argentina*. Oklahoma Museum of Natural History, Norman.

7   Barquez, R. M., M. A. Mares, and R. A. Ojeda. 1991. *Mammals of Tucuman*. Oklahoma Museum of Natural History, Norman.

8   Bates, P. J. J., and D. L. Harrison. 1997. *Bats of the Indian Subcontinent*. Harrison Zoological Museum, Kent, England.

9   Cairns, S. D., D. R. Calder, A. Brinckmann-Voss, C. B. Castro, P. R. Pugh, C. E. Cutress, W. C. Jaap, D. G. Fautin, R. J. Larson, G. R. Harbison, M. N. Arai, and D. M. Opresko. 1991. Common and scientific names of aquatic invertebrates from the United States and Canada: Cnidaria and Ctenophora. *American Fisheries Society Special Publication* 22:1–75.

10   Chakraborty, S. 1983. Contribution to the knowledge of the mammalian fauna of Jammu and Kashmir, India. *Records of the Zoological Survey of India* 38:1–29.

11   Chapman, J. A., and G. A. Feldhammer. 1982. *Wild mammals of North America: Biology, management, and economics*. Johns Hopkins University Press, Baltimore.

12   Chapman, J. A., and J. E. C. Flux, editors. 1990. *Rabbits, hares, and pikas: Status survey and conservation action plan*. IUCN, Gland, Switzerland.

13   Cole, F. R., D. M. Reeder, and D. E. Wilson. 1994. A synopsis of distribution patterns and the conservation of mammal species. *Journal of Mammalogy* 75:266–276.

14 Corbet, G. B. 1978. *The mammals of the Palearctic region: A taxonomic review.* British Museum (Natural History), London.

15 Corbet, G. B. 1984. *The mammals of the Palaearctic region: A taxonomic review: supplement.* British Museum (Natural History), London.

16 Corbet, G. B., and J. E. Hill. 1991. *A world list of mammalian species.* Oxford University Press, Oxford.

17 Corbet, G. B., and J. E. Hill. 1992. *Mammals of the Indomalayan region: A systematic review.* Oxford University Press, Oxford.

18 Corbet, G. B., and D. Ovenden. 1980. *The mammals of Britain and Europe.* Collins, London.

19 Corbet, G. B., and H. N. Southern, editors. 1977. *The handbook of British mammals.* Blackwell Scientific Publications, Oxford.

20 Cranbrook, Earl of. 1991. *Mammals of south-east Asia.* Oxford University Press, Singapore.

21 Darlington, P. J. 1957. *Zoogeography: The geographical distribution of animals.* John Wiley & Sons, New York.

22 Dorst, J., and P. Dandelot. 1989. *The mammals of Africa.* The Eaton Press, Norwalk, Connecticut.

23 East, R., editor. 1988. *Antelopes: Global survey and regional action plans. Part 1. East and northeast Africa.* IUCN, Gland, Switzerland.

24 Eisenberg, J. F. 1989. *Mammals of the Neotropics,* volume I. University of Chicago Press, Chicago.

25 Ellerman, J. R, and T. C. S. Morrison-Scott. 1951. *Checklist of Palearctic and Indian Mammals, 1758–1946.* British Museum (Natural History), London.

26 Emmons, L. H. 1997. *Neotropical rainforest mammals: A field guide,* second edition. University of Chicago Press, Chicago.

27 Fenton, M. B. 1992. *Bats.* Facts On File Publications, New York.

28 Flannery, T. 1995a. *Mammals of New Guinea.* Cornell University Press, Ithaca, NY.

29 Flannery, T. 1995b. *Mammals of the Southwest Pacific and Molucca Islands.* Cornell University Press, Ithaca, NY.

30 Ginsberg, J. R., and D. W. Macdonald, editors. 1990. *Foxes, wolves, jackals, and dogs: An action plan for the conservation of canids.* IUCN, Gland, Switzerland.

31 Goldman, E. A. 1918. The rice rats of North America (genus *Oryzomys*). *North American Fauna* 43:1–100.

32 Gonzalez, G. C., and C. G. Leal. 1984. *Mamiferos Silvestres de la Cuena de Mexico.* Programme on Man and the Biosphere and Instituto de Ecologia y Museo de Historia Natural de la Ciudad de Mexico.

33 Goodwin, G. G. 1946. Mammals of Costa Rica. *Bulletin of the American Museum of Natural History* 87:271–471.

34 Goodwin, G. G., and A. M. Greenhall. 1961. A review of the bats of Trinidad and Tobago. *Bulletin of the American Museum of Natural History* 122:187–302.

35 Gromov, I. M., and G. I. Baranova (eds.) 1981. *Katalog Mlekopitayushchikh SSSR: Catalog of mammals of the USSR.* Nauka, Leningrad (in Russian).

36   Gromov, I. M., and I. Y. Polyakov. 1977. *Polevki (Microtinae)*. Fauna SSSR: Mleko-pitayushchie, volume III, no. 8. Akademiya Nauk SSSR, Zoologicheskii Institut, New Series, no. 116, Nauka Publishers, Leningrad (in Russian).

37   Groombridge, B., editor. 1992. *Global biodiversity: Status of the Earth's living resources*. Chapman and Hall, London.

38   Hall, E. R. 1981. *The mammals of North America*, volumes I and II. John Wiley and Sons, New York, New York.

39   Hall, E. R., and K. R. Kelson. 1959. *The mammals of North America*, volume I. Ronald Press Company, New York.

40   Handley, C. O. Jr. 1966. Checklist of the Mammals of Panama, pp. 753–795 in *Ectoparasites of Panama* (R. L. Wenzel and V. J. Tipton, eds). Field Museum of Natural History, Chicago.

41   Happold, D. C. D. 1987. *The mammals of Nigeria*. Clarendon Press, Oxford.

42   Harcourt, C., and J. Thornback, editors. 1990. *Lemurs of Madagascar and the Comoros*. The IUCN Red Data Book. IUCN, Gland, Switzerland.

43   Harrison, D. L., and P. J. J. Bates. 1991. *The mammals of Arabia*, second edition. Harrison Zoological Museum, Kent, England.

44   Hayman, R. W., and J. E. Hill. 1971. *Order Chiroptera*, part 2, pp 1–73, in *The mammals of Africa: An identification manual*. Smithsonian Institution Press, Washington DC.

45   Heaney, L. R. 1985. Systematics of Oriental Pygmy Squirrels of the genera *Exilisciurus* and *Nannosciurus* (Mammalia: Sciuridae). *Miscellaneous Publications, Museum of Zoology, University of Michigan* 170:1–58.

46   Heaney, L. R., and R. C. B. Utzurrum. 1991. A review of the conservation status of Philippine land mammals. *ASBP Communications* 3:1–13.

47   Heaney, L. R., P. C. Gonzales, and A. C. Acala. 1987. An annotated checklist of the taxonomic and conservation status of land mammals in the Philippines. *Silliman Journal* 34:32–66.

48   Heaney, L. R., D. S. Balete, M. L. Dolar, A. C. Alcala, A. T. L. Dans, P. C. Gonzales, N. R. Ingle, M. V. Lepiten, W. L. R. Oliver, P. S. Ong, E. A. Rickart, B. R. Tabaranza, Jr., and R. C. B. Utzurrum. 1998. A synopsis of the mammalian fauna of the Philippine Islands. *Fieldiana: Zoology*, new series, no. 88:1–61.

49   Heller, E. 1910. New species of Insectivores from British East Africa, Uganda, and the Sudan. *Smithsonian Miscellaneous Collections* 56:1–8.

50   Hershkovitz, P. 1972. The Recent mammals of the Neotropical Region: A zoogeographic and ecological review, pp. 311–431, in *Evolution, mammals, and southern continents* (A. Keast, F. C. Erk, and B. Glass, eds.). State University of New York, Albany.

51   Hoffmann, R. S. 1996. Noteworthy shrews and voles from the Xizang-Qinghai Plateau, pp. 155–168, in *Contributions in mammalogy: A memorial volume honoring Dr. J. Knox Jones, Jr.* (H. H. Genoways and R. J. Baker, eds.). Museum of Texas Tech University, Lubbock, Texas.

52   Hoffmeister, D. F. 1986. *Mammals of Arizona*. University of Arizona Press, Tucson.

53   Howard, R. and A. Moore. 1994. *A complete checklist of the birds of the world*, second edition. Academic Press, New York

54   Jones, C., R. S. Hoffmann, D. W. Rice, M. D. Engstrom, R. D. Bradley, D. J. Schmidly, C. A. Jones, and R. J. Baker. 1997. Revised checklist of North American mammals north of Mexico, 1997. *Occasional Papers, The Museum, Texas Tech University* 173:1–19.

55   Kennedy, M., editor. 1992. *Australasian marsupials and monotremes: An action plan for their conservation.* IUCN, Gland, Switzerland.

56   Kingdon, J. 1971. *East African mammals: An atlas of evolution in Africa*, volume I. Academic Press, London.

57   Kingdon, J. 1974. *East African mammals: An atlas of evolution in Africa*, volume IIa. *Insectivores and bats.* Academic Press, London.

58   Kingdon, J. 1974. *East African mammals: An atlas of evolution in Africa*, volume IIb. *Hares and rodents.* Academic Press, London.

59   Kingdon, J. 1977. *East African mammals: An atlas of evolution in Africa*, volume IIIa. *Carnivores.* Academic Press, London.

60   Kingdon, J. 1979. *East African mammals: An atlas of evolution in Africa*, volume IIIb. *Large mammals.* Academic Press, London.

61   Kingdon, J. 1982. *East African mammals: An atlas of evolution in Africa*, volume IIIc, d. *Bovids.* Academic Press, London.

62   Kingdon, J. 1991. *Arabian mammals.* Academic Press, London.

63   Kingdon, J. 1997. *The Kingdon field guide to African mammals.* Academic Press, London.

64   Kleiman, D. G. 1975. *The biology and conservation of the Callitrichidae.* Conservation and Research Center, National Zoological Park, Smithsonian Institution, Smithsonian Institution Press, Washington, DC.

65   Klinowska, M., editor. 1991. *Dolphins, porpoises, and whales of the world.* The IUCN Red Data Book. IUCN, Gland, Switzerland.

66   Lee, P. C., J. Thornback, and E. L. Bennett, editors. 1988. *Threatened primates of Africa.* The IUCN Red Data Book. IUCN, Gland, Switzerland.

67   Lekagul, B., and J. A. McNeely. 1977. *Mammals of Thailand.* Association for the Conservation of Wildlife, Sahakarnbhat Co., Bangkok.

68   Lin, L. K., Lee, L. L., and Cheng, H. C. 1997. *Bats of Taiwan.* National Taiwan University, Taipei.

69   Macdonald, D., editor. 1984. *The encyclopedia of mammals*, 2 volumes. Allen & Unwin, London.

70   Mares, M. A., R. A. Ojeda, and R. M. Barquez. 1989. *Guide to the mammals of Salta Province, Argentina.* University of Oklahoma Press, Norman.

71   Marshall, J. T. 1986. Systematics of the genus *Mus. Current topics in microbiology and immunology* 127:13–18.

72   Matson, J. O., and R. H. Baker. 1986. Mammals of Zacatecas. *Special Publications, The Museum, Texas Tech University* 24:1–88.

73   Mearns, E. A. 1905. Descriptions of new genera and species of mammals from the Philippine Islands. *Proceedings of the U.S. National Museum* 28:425–460.

74   Medway, Lord. 1977. *Mammals of Borneo: Field keys and an annotated checklist.* Malaysian Branch of the Royal Asiatic Society.

75   Mickleburgh, S. P., A. M. Hutson, and P. A. Racey, editors. 1992. *Old world fruit bats: An action plan for their conservation.* IUCN, Gland, Switzerland.

76   Mittermeier, R. A., I. Tattersall, W. R. Konstant, D. M. Meyers, and S. D. Nash. 1994. *Lemurs of Madagascar.* Conservation International, Washington, DC.

77   Miller, J. K., editor. 1992 *The common names of North American butterflies.* Smithsonian Institution Press, Washington, DC.

78   Mills, G., and L. Hes. 1997. *The complete book of Southern African mammals.* Struik Winchester, Capetown, South Africa.

79   Mochi, U., and T. D. Carter. 1971. *Hoofed mammals of the world.* Charles Scribner's Sons, New York.

80   Monroe, B. L., and C. G. Sibley. 1993. *A world checklist of birds.* Yale University Press, New Haven, CT.

81   Morgan, G. S. 1989. *Geocapromys thoracatus. Mammalian Species* 341:1–5.

82   Musser, G. G. 1987. The mammals of Sulawesi, pp. 73–93, in *Biogeographical evolution of the Malay Archipelago* (T. C. Whitmore, ed.). Oxford University Press, Oxford.

83   Musser, G. G. 1981. The giant rat of Flores and its relatives east of Borneo and Bali. *Bulletin of the American Museum of Natural History* 169:71–175.

84   Nicoll, M. E., and G. B. Rathbun, editors. 1990. *African Insectivora and elephant-shrews: An action plan for their conservation.* IUCN, Gland, Switzerland.

85   Niethammer, J., and F. Krapp, editors. 1990. *Handbuch der Säugetiere Europas, 3/I.* Aula-Verlag, Wiesbaden.

86   Nowak, R. M. 1999. *Walker's mammals of the world,* 6th edition. Volumes I and II. The Johns Hopkins University Press, Baltimore.

87   Ognev, S. I. 1962–4. *Mammals of eastern Europe and northern Asia.* Israel Program for Scientific Translations, Jerusalem.

88   Oliver, W. L. R., editor. 1993. *Pigs, peccaries, and hippos: Status survey and conservation action plan.* IUCN, Gland, Switzerland.

89   Osgood, W. H. 1912. Mammals from western Venezuela and eastern Columbia. *Field Museum of Natural History, Zoological Series* 10:32–67.

90   Osgood, W. H. 1914. Mammals of an expedition across northern Peru. *Fieldiana: Zoology* 10:143–185.

91   Payne, J., C. M. Francis, and K. Phillipps. 1985. *A field guide to the mammals of Borneo.* Sabah Society with World Wildlife Fund Malaysia, Kuala Lumpur.

92   Phillips, W. W. A. 1980–1984. Manual of the mammals of Sri Lanka. Second revised ed. *Wildlife and Nature Protection Society of Sri Lanka* 1(1980): 1–116; 2(1980): 117–267; 3(1984): 268–389.

93   Phillips, W. W. A. 1932. Additions to the fauna of Ceylon, no. 1: Two new rodents

from the hills of central Ceylon. *Ceylon Journal of Science, Section B. Zoology & Geology, Spolia Zeylonica* 16:323–327.

94   Qumsiyeh, M. B., and J. K. Jones, Jr. 1986. *Rhinopoma hardwickii* and *Rhinopoma muscatellum. Mammalian Species,* 263:1–5.

95   Ralls, K., and R. L. Brownell, Jr. 1991. A whale of a new species. *Nature* 350:560.

96   Ramirez-Pulido, J., R. Lopez-Wilchis, C. Mudespacher Ziehl, and I. J. Lira. 1982. *Catalogo de los Mamíferos terrestres nativas de Mexico.* Universidad Autonoma Metropolitana-Iztapalapa, Mexico City.

97   Redford, K. H., and J. F. Eisenberg. 1992. *Mammals of the Neotropics,* volume II. University of Chicago Press, Chicago, Illinois.

98   Reid, F. A. 1997. *A field guide to the mammals of Central America and Southeast Mexico.* Oxford University Press, New York.

99   Reijnders, P., S. Brasseur, J. van der Toorn, P. van der Wolf, I. Boyd, J. Harwood, editors. 1993. *Seals, fur seals, sea lions, and walrus: Status survey and conservation action plan.* IUCN, Gland, Switzerland.

100   Rhoads, S. N. 1896. Mammals collected by Dr. A. Donaldson Smith during his expedition to Lake Rudolf. *Proceedings of the Academy of Natural Sciences of Philadelphia* 1896:517–546.

101   Roberts, A. 1951. *The mammals of South Africa.* Johannesburg, Trustees of "The mammals of South Africa" book fund.

102   Roberts, T. J. 1997. *The mammals of Pakistan,* revised edition. Oxford University Press, Oxford.

103   Rosevear, D. R. 1965. *The bats of West Africa.* British Museum (Natural History), London.

104   Rosevear, D. R. 1969. *The rodents of West Africa.* British Museum (Natural History), London.

105   Rylands, A. B., R. A. Mittermeier, and E. R. Luna. 1995. A species list for the New World primates (Platyrrhini): Distribution by country, endemism, and conservation status according to the Mace-Land system. *Neotropical Primates* 3(suppl): 113–160.

106   Schreiber, A., R. Wirth, M. Riffel, and H. V. Rompaey, editors. 1989. *Weasels, civets, mongooses, and their relatives: An action plan for the conservation of mustelids and viverrids.* IUCN, Gland, Switzerland.

107   Skinner, J. D., and R. H. N. Smithers. 1990. *The mammals of the southern African subregion.* University of Pretoria, Pretoria, South Africa.

108   Smithers, R. H. N. 1983. *The mammals of the southern African subregion.* University of Pretoria, Pretoria, South Africa.

109   Sokolov, W. E. 1984. *A dictionary of animal names in five languages.* Russky Yazyk, Moscow.

110   Strahan, R., editor. 1995. *Mammals of Australia.* Smithsonian Institution Press, Washington, DC.

111   Topachevskii, V. A. 1976. *Fauna of the USSR: Mammals. Mole rats, Spalacidae.* Amerind Publishing Co, New Delhi.

112   Wilson, D. E. 1983. Checklist of mammals, pp. 443–447, in *Costa Rica Natural History* (D. H. Janzen, ed.). University of Chicago Press, Chicago.

113   Wilson, D. E., and D. M. Reeder (eds.). 1993. *Mammal species of the world*, 2nd edition. Smithsonian Institution Press, Washington, DC.

114   Wilson, D. E., and S. Ruff. 1999. *The Smithsonian book of North American mammals*. Smithsonian Institution Press, Washington, DC.

115   Yin, U.T. 1993. *Wild Mammals of Myanmar*. Forest Department, Myanmar, Yangon.

116   Zhang, Y. Z. 1997. *Distribution of mammalian species in China*. China Forestry Publishing House, Beijing.

117   Ziegler, A. C. 1971. Suggested vernacular names for New Guinea recent land mammals. *Occasional Papers of Bernice P. Bishop Museum* 24:140–153.

118   Ziegler, A. C. 1982. An ecological check-list of New Guinea recent mammals, pp. 863–894, in *Biogeography and ecology of New Guinea*, volume 2 (J. L. Gressitt, ed.). Monographiae Biologicae, volume 42. W. Junk, The Hague.

INDEX TO SCIENTIFIC NAMES

This index includes the scientific names for all the orders, familes, subfamilies, and genera listed in the text. It does not include specific epithets.

Abditomys, 125
Abrawayaomys, 146
Abrocoma, 167
Abrocomidae, 167
Acerodon, 35
Acinonychinae, 76
Acinonyx, 76
Acomys, 125
Aconaemys, 166
Acrobates, 18
Acrobatidae, 18
Addax, 99
Aepeomys, 146
Aepyceros, 95
Aepycerotinae, 95
Aepyprymnus, 15
Aeretes, 108
Aeromys, 108
Aethalops, 35
Aethomys, 125–126
Agouti, 165
Agoutidae, 165
Ailurinae, 84
Ailuropoda, 84
Ailurops, 14
Ailurus, 84
Akodon, 146–147
Alcelaphus, 95
Alces, 94
Alcetaphinae, 95
Alionycteris, 35
Allactaga, 113
Allactaginae, 113–114

Allactodipus, 113
Allenopithecus, 72
Allocebus, 66
Allocricetulus, 120
Alopex, 75
Alouatta, 70
Alouattinae, 70
Alticola, 115
Amblonyx, 80
Amblyrhiza, 171
Amblysomus, 21–22
Ametrida, 51
Ammodillus, 122
Ammodorcas, 95
Ammospermophilus, 100
Ammotragus, 98
Amorphochilus, 54
Anathana, 34
Andalgalomys, 147
Andinomys, 147
Anisomys, 126
Anomaluridae, 160–161
Anomalurinae, 160
Anomalurus, 160
Anonymomys, 126
Anotomys, 148
Anoura, 50
Anourosorex, 29
Antechinus, 11
Anthops, 45
Antidorcas, 95
Antilocapra, 95
Antilocapridae, 95

Antilope, 96
Antilopinae, 95–97
Antrozous, 55
Aonyx, 80
Aotinae, 70
Aotus, 70
Aplodontia, 100
Aplodontidae, 100
Apodemus, 126
Apomys, 126–127
Aproteles, 35
Arborimus, 115
Archboldomys, 127
Arctictis, 85
Arctocebus, 68
Arctocephalus, 82
Arctogalidia, 85
Arctonyx, 80
Ardops, 51
Ariteus, 51
Artibeus, 51–52
Artiodactyla, 92–100
Arvicanthis, 127
Arvicola, 116
Arvicolinae, 115–119
Asellia, 46
Aselliscus, 46
Atelerix, 22
Ateles, 70–71
Atelinae, 70–71
Atelocynus, 75
Atherurus, 162
Atilax, 78

Atlantoxerus, 101
Auliscomys, 148
Australophocaena, 88
Avahi, 67
Axis, 93

Babyrousa, 92
Babyrousinae, 92
Baiomys, 148
Balaena, 86
Balaenidae, 86
Balaenoptera, 86–87
Balaenopteridae, 86–87
Balantiopteryx, 41
Balionycteris, 35
Bandicota, 127
Barbastella, 55
Bassaricyon, 84
Bassariscus, 84
Bathyergidae, 162
Bathyergus, 162
Batomys, 127
Bdeogale, 78
Beamys, 121
Belomys, 108
Berardius, 89
Berylmys, 127
Bettongia, 15
Bibimys, 148
Bison, 97
Biswamoyopterus, 108
Blanfordimys, 116
Blarina, 29
Blarinella, 29
Blarinomys, 148
Blastocerus, 94
Bolomys, 148
Boneia, 35
Boromys, 169
Bos, 97
Boselaphus, 97
Bovidae, 95–100
Bovinae, 97
Brachiones, 122
Brachylagus, 172
Brachyphylla, 50
Brachyphyllinae, 50
Brachytarsomys, 144
Brachyteles, 71
Brachyuromys, 144
Bradypodidae, 18–19
Bradypus, 18–19

Brotomys, 169
Bubalus, 97
Budorcas, 98
Bullimus, 127
Bunolagus, 172
Bunomys, 127–128
Burramyidae, 17
Burramys, 17

Cabassous, 19
Cacajao, 72
Caenolestes, 10
Caenolestidae, 10
Calcochloris, 22
Callicebinae, 71
Callicebus, 71
Callimico, 69
Callithrix, 69
Callitrichidae, 69–70
Callorhinus, 82
Callosciurus, 101
Calomys, 148
Calomyscinae, 120
Calomyscus, 120
Caloprymnus, 15
Caluromyinae, 7
Caluromys, 7
Caluromysiops, 7
Camelidae, 93
Camelus, 93
Canariomys, 128
Canidae, 75–76
Canis, 75
Cannomys, 146
Canusomys, 120
Caperea, 87
Capra, 98
Capreolinae, 94–95
Capreolus, 94
Caprinae, 98–99
Caprolagus, 172
Capromyidae, 170
Capromyinae, 170
Capromys, 170
Caracal, 77
Cardiocraniinae, 114
Cardiocranius, 114
Cardioderma, 43
Carnivora, 75–86
Carollia, 51
Carolliinae, 51
Carpomys, 128

Carterodon, 168
Casinycteris, 35
Castor, 110
Castoridae, 110
Catagonus, 92
Catopuma, 77
Cavia, 164
Caviidae, 164
Caviinae, 164
Cebidae, 70–72
Cebinae, 71–72
Cebus, 71
Celaenomys, 128
Centronycteris, 41
Centurio, 52
Cephalophinae, 99
Cephalophus, 99
Cephalorhynchus, 87
Ceratotherium, 91
Cercartetus, 17
Cercocebus, 72
Cercopithecidae, 72–74
Cercopithecinae, 72–73
Cercopithecus, 72–73
Cerdocyon, 76
Cervidae, 93–95
Cervinae, 93–94
Cervus, 93–94
Cetacea, 86–90
Chaerephon, 64
Chaeropus, 13
Chaetodipus, 112–113
Chaetomyinae, 167
Chaetomys, 167
Chaetophractus, 19
Chalinolobus, 55
Cheirogaleidae, 66–67
Cheirogaleus, 66
Cheirogalinae, 66–67
Cheiromeles, 64
Chelemys, 148
Chibchanomys, 149
Chilomys, 149
Chimarrogale, 29
Chinchilla, 164
Chinchillidae, 164
Chinchillula, 149
Chionomys, 116
Chiroderma, 52
Chiromyscus, 128
Chironax, 35
Chironectes, 7

Chiropodomys, 128
Chiropotes, 72
Chiroptera, 35–66
Chiruromys, 128
Chlamyphorinae, 19
Chlamyphorus, 19
Chlorocebus, 73
Chlorotalpa, 22
Choeroniscus, 50
Choeronycteris, 50
Choloepus, 19
Chroeomys, 149
Chrotogale, 85
Chrotomys, 128
Chrotopterus, 48
Chrysochloridae, 21–22
Chrysochloris, 22
Chrysocyon, 76
Chrysospalax, 22
Civettictis, 86
Clethrionomys, 116
Clidomyinae, 170
Clidomys, 170
Cloeotis, 46
Clyomys, 168
Coccymys, 128
Coelops, 46
Coendou, 163
Coleura, 41
Colobinae, 73–74
Colobus, 73–74
Colomys, 128
Comura, 42
Condylura, 32
Conepatus, 81
Congosorex, 23
Conilurus, 129
Connochaetes, 95
Coryphomys, 129
Craseonycteridae, 41
Craseonycteris, 41
Crateromys, 129
Cremnomys, 129
Cricetinae, 120
Cricetomyinae, 121
Cricetomys, 121
Cricetulus, 120
Cricetus, 120
Crocidura, 23–27
Crocidurinae, 23–29
Crocuta, 79
Crossarchus, 78

Crossomys, 129
Crunomys, 129
Cryptochloris, 22
Cryptomys, 162
Cryptoprocta, 85
Cryptoproctinae, 85
Cryptotis, 29–30
Ctenodactylidae, 161
Ctenodactylus, 161
Ctenomyidae, 165–166
Ctenomys, 165–166
Cuon, 76
Cyclopes, 20
Cynictis, 78
Cynocephalidae, 35
Cynocephalus, 35
Cynogale, 85
Cynomys, 101
Cynopterus, 35
Cystophora, 83
Cyttarops, 42

Dacnomys, 129
Dactylomyinae, 167
Dactylomys, 167
Dactylopsila, 18
Dama, 94
Damaliscus, 95
Dasycercus, 11
Dasykaluta, 11
Dasymys, 129
Dasypodidae, 19–20
Dasypodinae, 19–20
Dasyprocta, 165
Dasyproctidae, 165
Dasypus, 19
Dasyuridae, 11–13
Dasyuromorphia, 11–13
Dasyurus, 11
Daubentonia, 68
Daubentoniidae, 68
Delanymys, 145
Delomys, 149
Delphinapterus, 88
Delphinidae, 87–88
Delphinus, 87
Dendrogale, 34
Dendrohyrax, 91
Dendrolagus, 15–16
Dendromurinae, 121–122
Dendromus, 121
Dendroprionomys, 121

Deomys, 121
Dephomys, 129
Dermoptera, 35
Desmana, 32
Desmaninae, 32
Desmodilliscus, 122
Desmodillus, 122
Desmodontinae, 53–54
Desmodus, 53
Desmomys, 129
Diaemus, 54
Dicerorhinus, 91
Diceros, 91
Diclidurus, 42
Dicrostonyx, 116
Didelphinae, 7–10
Didelphidae, 7–10
Didelphimorphia, 7–10
Didelphis, 7–8
Dinaromys, 116
Dinomyidae, 164
Dinomys, 164
Diomys, 129
Diphylla, 54
Diplogale, 85
Diplomesodon, 27
Diplomys, 167
Diplothrix, 129
Dipodidae, 113–115
Dipodinae, 114
Dipodomyinae, 111–112
Dipodomys, 111–112
Diprotodontia, 14–18
Dipus, 114
Distoechurus, 18
Dobsonia, 36
Dolichotinae, 164
Dolichotis, 164
Dologale, 78
Dorcatragus, 96
Dorcopsis, 16
Dorcopsulus, 16
Dremomys, 101
Dromiciops, 10
Dryomys, 161
Dugong, 90
Dugongidae, 90
Dusicyon, 76
Dyacopterus, 36

Echimyidae, 167–169
Echimyinae, 167–168

Echimys, 167–168
Echinoprocta, 163
Echinops, 21
Echinosorex, 23
Echiothrix, 130
Echymipera, 13
Ectophylla, 52
Eidolon, 36
Eira, 81
Elaphodus, 94
Elaphurus, 94
Elasmodontomys, 171
Elephantidae, 90
Elephantulus, 173–174
Elephas, 90
Eligmodontia, 149
Eliomys, 161
Eliurus, 144–145
Ellobius, 116–117
Emballonura, 42
Emballonuridae, 41–43
Enhydra, 80
Eolagurus, 117
Eonycteris, 41
Eothenomys, 117
Eozapus, 115
Epixerus, 101
Epomophorus, 36
Epomops, 37
Eptesicus, 55–56
Equidae, 90–91
Equus, 90–91
Eremitalpa, 22
Eremodipus, 114
Erethizon, 163
Erethizontidae, 163–164
Erignathus, 83
Erinaceidae, 22–23
Erinaceinae, 22–23
Erinaceus, 22
Eropeplus, 130
Erophylla, 50
Erythrocebus, 73
Eschrichtiidae, 87
Eschrichtius, 87
Eubalaena, 86
Euchoreutes, 114
Euchoreutinae, 114
Euderma, 56
Eudiscopus, 56
Eulemur, 67
Eumetopias, 83

Eumops, 64
Eumysopinae, 168–169
Euneomys, 149
Euoticus, 68
Eupetaurus, 108
Euphractus, 19
Eupleres, 85
Euplerinae, 85
Euroscaptor, 32–33
Euryzygomatomys, 168
Exilisciurus, 101–102

Felidae, 76–78
Felinae, 77–78
Felis, 77
Felovia, 161
Feresa, 87
Feroculus, 27
Fossa, 85
Funambulus, 102
Funisciurus, 102
Furipteridae, 54
Furipterus, 54

Galago, 68
Galagoides, 68
Galagonidae, 68
Galea, 164
Galemys, 32
Galenomys, 149
Galerella, 79
Galictis, 81
Galidia, 78
Galidictis, 78
Galidiinae, 78
Gazella, 96
Genetta, 86
Geocapromys, 170
Geogale, 20
Geogalinae, 20
Geomyidae, 110–111
Geomys, 110
Georychus, 162
Geoxus, 149
Gerbillinae, 122–125
Gerbillurus, 122
Gerbillus, 122–124
Giraffa, 93
Giraffidae, 93
Glaucomys, 108
Glironia, 7
Glirulus, 162

Glischropus, 56
Globicephala, 87
Glossophaga, 50
Glossophaginae, 50–51
Glyphotes, 102
Golunda, 130
Gorilla, 75
Gracilinanus, 8
Grammomys, 130
Grampus, 87
Graomys, 149
Graphiurinae, 161
Graphiurus, 161
Gulo, 81
Gymnobelideus, 18
Gymnuromys, 145

Habromys, 149
Hadromys, 130
Haeromys, 130
Halichoerus, 83
Hapalemur, 67
Hapalomys, 130
Haplonycteris, 37
Harpiocephalus, 63
Harpyionycteris, 37
Heimyscus, 130
Helarctos, 84
Heliophobius, 162
Heliosciurus, 102
Helogale, 79
Hemibelideus, 17
Hemicentetes, 21
Hemiechinus, 22–23
Hemigalinae, 85
Hemigalus, 85
Hemitragus, 98
Heptaxodontidae, 170–171
Heptaxodontinae, 171
Herpailurus, 77
Herpestes, 79
Herpestidae, 78–79
Herpestinae, 78–79
Hesperoptenus, 56–57
Heterocephalus, 162
Heterohyrax, 91
Heteromyidae, 111–113
Heteromyinae, 112
Heteromys, 112
Heteropsomyinae, 169
Heteropsomys, 169
Hexaprotodon, 92

Hexolobodon, 170
Hexolobodontinae, 170
Hippocamelus, 94
Hippopotamidae, 92
Hippopotamus, 92–93
Hipposiderinae, 45–48
Hipposideros, 46–47
Hippotraginae, 99
Hippotragus, 99
Histiotus, 57
Hodomys, 149
Holochilus, 150
Hominidae, 75
Homo, 75
Hoplomys, 168
Hyaena, 79
Hyaenidae, 79–80
Hyaeninae, 79–80
Hybomys, 130
Hydrochaeridae, 165
Hydrochaeris, 165
Hydrodamalis, 90
Hydromys, 130–131
Hydropotes, 94
Hydropotinae, 94
Hydrurga, 83
Hyemoschus, 93
Hylobates, 75
Hylobatidae, 75
Hylochoerus, 92
Hylomyinae, 23
Hylomys, 23
Hylomyscus, 131
Hylonycteris, 50
Hylopetes, 108–109
Hyomys, 131
Hyosciurus, 102
Hyperacrius, 117
Hyperoodon, 89
Hypogeomys, 145
Hypsignathus, 37
Hypsiprymnodon, 15
Hyracoidea, 91
Hystricidae, 162–163
Hystrix, 163

Ia, 57
Ichneumia, 79
Ichthyomys, 150
Ictonyx, 81
Idionycteris, 57
Idiurus, 160

Indopacetus, 89
Indri, 67
Indridae, 67–68
Inia, 89
Insectivora, 20–34
Iomys, 109
Irenomys, 150
Isolobodon, 170
Isolobodontinae, 170
Isoodon, 13
Isothrix, 168
Isthmomys, 150

Jaculus, 114
Juscelinomys, 150

Kadarsanomys, 131
Kannabateomys, 167
Kerivoula, 54–55
Kerivoulinae, 54–55
Kerodon, 164
Kobus, 100
Kogia, 89
Komodomys, 131
Kunsia, 150

Laephotis, 57
Lagenodelphis, 87
Lagenorhynchus, 87
Lagidium, 164
Lagomorpha, 171–173
Lagorchestes, 16
Lagostomus, 164
Lagostrophus, 16
Lagothrix, 71
Lagurus, 117
Lama, 93
Lamottemys, 131
Lariscus, 102–103
Lasionycteris, 57
Lasiopodomys, 117
Lasiorhinus, 14
Lasiurus, 57
Latidens, 37
Lavia, 43
Leggadina, 131
Leimacomys, 121
Leithiinae, 161–162
Lemmiscus, 117
Lemmus, 117
Lemniscomys, 131–132
Lemur, 67

Lemuridae, 67
Lenomys, 132
Lenothrix, 132
Lenoxus, 150
Leontopithecus, 69
Leopardus, 77
Leopoldamys, 132
Lepilemur, 67
Leporidae, 172–173
Leporillus, 132
Leptailurus, 77
Leptomys, 132
Leptonychotes, 83
Leptonycteris, 51
Lepus, 172
Lestodelphys, 8
Lestoros, 10
Liberiictis, 79
Lichonycteris, 51
Limnogale, 21
Limnomys, 132
Liomys, 112
Lionycteris, 49
Lipotes, 89
Lissodelphis, 88
Litocranius, 96
Lobodon, 83
Lonchophylla, 49–50
Lonchophyllinae, 49–50
Lonchorhina, 48
Lonchothrix, 168
Lontra, 80
Lophiomyinae, 125
Lophiomys, 125
Lophocebus, 73
Lophuromys, 132–133
Lorentzimys, 133
Loridae, 68
Loris, 68
Loxodonta, 90
Lutra, 80
Lutreolina, 8
Lutrinae, 80
Lutrogale, 80
Lycaon, 76
Lyncodon, 81
Lynx, 77

Macaca, 73
Macroderma, 43
Macrogalidia, 85
Macroglossinae, 41

Macroglossus, 41
Macrophyllum, 48
Macropodidae, 15–17
Macropus, 16
Macroscelidea, 173–174
Macroscelididae, 173–174
Macrotarsomys, 145
Macrotis, 13
Macrotus, 48
Macruromys, 133
Madoqua, 96
Makalata, 168
Malacomys, 133
Malacothrix, 121
Mallomys, 133
Malpaisomys, 133
Mandrillus, 73
Manidae, 100
Manis, 100
Margaretamys, 133
Marmosa, 8
Marmosops, 8–9
Marmota, 103
Martes, 81
Massoutiera, 161
Mastomys, 133–134
Maxomys, 134
Mayermys, 134
Mazama, 94–95
Megadendromus, 121
Megaderma, 43
Megadermatidae, 43
Megadontomys, 150
Megaerops, 37
Megaglossus, 41
Megaladapidae, 67
Megalomys, 150
Megalonychidae, 19
Megaptera, 87
Megasorex, 30
Melanomys, 150
Melasmothrix, 134
Meles, 80
Melinae, 80
Mellivora, 81
Mellivorinae, 81
Melogale, 80
Melomys, 134–135
Melonycteris, 41
Melursus, 84
Menetes, 103
Mephitinae, 81

Mephitis, 81
Meriones, 124
Mesechinus, 23
Mesembriomys, 135
Mesocapromys, 170
Mesocricetus, 120
Mesomys, 168
Mesophylla, 52
Mesoplodon, 89–90
Metachirus, 9
Micoureus, 9
Microbiotheria, 10
Microbiotheriidae, 10
Microcavia, 164
Microcebus, 67
Microdillus, 124
Microdipodops, 112
Microgale, 21
Microhydromys, 135
Micromys, 135
Micronycteris, 48–49
Microperoryctes, 13
Micropotamogale, 21
Micropteropus, 37
Microryzomys, 150
Microsciurus, 103
Microtus, 117–119
Millardia, 135
Mimetillus, 57
Mimon, 49
Miniopterinae, 63
Miniopterus, 63
Miopithecus, 73
Mirounga, 83
Mogera, 33
Molossidae, 64–66
Molossops, 64–65
Molossus, 65s
Monachus, 83
Monodelphis, 9–10
Monodon, 88
Monodontidae, 88
Monophyllus, 51
Monotremata, 7
Mops, 65
Mormoopidae, 48
Mormoops, 48
Mormopterus, 65
Moschidae, 93
Moschiola, 93
Moschus, 93
Mosia, 42

Mungos, 79
Mungotictis, 78
Muntiacinae, 94
Muntiacus, 94
Murexia, 11
Muriculus, 135
Muridae, 115–160
Murina, 63
Murinae, 125–144
Murininae, 63
Mus, 135–136
Muscardinus, 162
Musonycteris, 51
Mustela, 82
Mustelidae, 80–82
Mustelinae, 81–82
Mydaus, 80
Mylomys, 136
Myocastor, 171
Myocastoridae, 171
Myoictis, 11
Myomimus, 162
Myomys, 136–137
Myonycteris, 37
Myoprocta, 165
Myopterus, 65
Myopus, 119
Myosciurus, 103
Myosorex, 27
Myospalacinae, 144
Myospalax, 144
Myotis, 57–59
Myoxidae, 161–162
Myoxinae, 162
Myoxus, 162
Myrmecobiidae, 11
Myrmecobius, 11
Myrmecophaga, 20
Myrmecophagidae, 20
Mysateles, 170
Mystacina, 64
Mystacinidae, 64
Mystromyinae, 144
Mystromys, 144
Myzopoda, 54
Myzopodidae, 54

Naemorhedus, 98
Nandinia, 85
Nandiniinae, 85
Nannosciurus, 103
Nannospalax, 160

Nanonycteris, 37
Napaeozapus, 115
Nasalis, 74
Nasua, 84
Nasuella, 84
Natalidae, 54
Natalus, 54
Neacomys, 151
Nectogale, 30
Nectomys, 151
Nelsonia, 151
Neobalaenidae, 87
Neofelis, 78
Neofiber, 119
Neohydromys, 137
Neomys, 30
Neophascogale, 11
Neophoca, 83
Neophocaena, 89
Neopteryx, 37
Neotoma, 151
Neotomodon, 151
Neotomys, 151
Neotragus, 96
Nesokia, 137
Nesolagus, 173
Nesomyinae, 144–145
Nesomys, 145
Nesophontes, 20
Nesophontidae, 20
Nesoryzomys, 152
Nesoscaptor, 33
Neurotrichus, 33
Neusticomys, 152
Ningaui, 12
Niviventer, 137
Noctilio, 48
Noctilionidae, 48
Notiomys, 152
Notiosorex, 30
Notomys, 137
Notopteris, 41
Notoryctemorphia, 14
Notoryctes, 14
Notoryctidae, 14
Nyctalus, 59–60
Nyctereutes, 76
Nycteridae, 43
Nycteris, 43
Nycticebus, 68
Nycticeius, 60
Nyctimene, 37–38

Nyctinomops, 66
Nyctomys, 152
Nyctophilus, 60

Ochotona, 171
Ochotonidae, 171–172
Ochrotomys, 152
Octodon, 166
Octodontidae, 166–167
Octodontomys, 166
Octomys, 166
Odobenidae, 82
Odobenus, 82
Odocoileus, 95
Oecomys, 152
Oenomys, 138
Okapia, 93
Olallamys, 167
Oligoryzomys, 152–153
Ommatophoca, 83
Oncifelis, 77
Ondatra, 119
Onychogalea, 16
Onychomys, 153
Orcaella, 88
Orcinus, 88
Oreailurus, 77
Oreamnos, 98
Oreotragus, 96
Ornithorhynchidae, 7
Ornithorhynchus, 7
Orthogeomys, 110
Orycteropodidae, 91
Orycteropus, 91
Oryctolagus, 173
Oryx, 99
Oryzomys, 153–154
Oryzorictes, 21
Oryzorictinae, 21
Osbornictis, 86
Osgoodomys, 154
Otaria, 83
Otariidae, 82–83
Otocolobus, 77
Otocyon, 76
Otolemur, 68
Otomops, 66
Otomyinae, 145
Otomys, 145
Otonycteris, 60
Otonyctomys, 154
Otopteropus, 38

Ototylomys, 154
Ourebia, 96
Ovibos, 98
Ovis, 98
Oxymycterus, 154
Ozotoceros, 95

Pachyuromys, 124
Paguma, 85
Palawanomys, 138
Pan, 75
Panthera, 78
Pantherinae, 78
Pantholops, 96
Papagomys, 138
Papio, 73
Pappogeomys, 110–111
Paracoelops, 47
Paracrocidura, 28
Paracynictis, 79
Paradipodinae, 114
Paradipus, 114
Paradoxurinae, 85
Paradoxurus, 85
Parahyaena, 80
Parahydromys, 138
Paraleptomys, 138
Parantechinus, 12
Paranyctimene, 38
Parascalops, 33
Parascaptor, 33
Paraxerus, 103–104
Pardofelis, 78
Parotomys, 145
Paruromys, 138
Paucituberculata, 10
Paulamys, 138
Pecari, 92
Pectinator, 161
Pedetes, 161
Pedetidae, 161
Pelea, 100
Peleinae, 100
Pelomys, 138
Pentalagus, 173
Penthetor, 38
Peponocephala, 88
Peramelemorphia, 13–14
Perameles, 13
Peramelidae, 13
Perissodactyla, 90–91
Perodicticus, 68

Perognathinae, 112–113
Perognathus, 113
Peromyscus, 154–156
Peropteryx, 42
Peroryctes, 14
Peroryctidae, 13–14
Petauridae, 18
Petaurillus, 109
Petaurista, 109
Petauroides, 17
Petaurus, 18
Petinomys, 109
Petrogale, 17
Petromuridae, 163
Petromus, 163
Petromyscinae, 145
Petromyscus, 145
Petropseudes, 17
Phacochoerinae, 92
Phacochoerus, 92
Phaenomys, 156
Phalanger, 14
Phalangeridae, 14–15
Phaner, 67
Phanerinae, 67
Pharotis, 60
Phascogale, 12
Phascolarctidae, 14
Phascolarctos, 14
Phascolosorex, 12
Phaulomys, 119
Phenacomys, 119
Philander, 10
Philetor, 60
Phloeomys, 138–139
Phoca, 83
Phocarctos, 83
Phocidae, 83
Phocoena, 89
Phocoenidae, 88–89
Phocoenoides, 89
Phodopus, 120
Pholidota, 100
Phylloderma, 49
Phyllonycterinae, 50
Phyllonycteris, 50
Phyllops, 52
Phyllostomidae, 48–54
Phyllostominae, 48–49
Phyllostomus, 49
Phyllotis, 156
Physeter, 89

Physeteridae, 89
Pipistrellus, 60–61
Pithecheir, 139
Pithecia, 72
Pitheciinae, 72
Plagiodontia, 170
Plagiodontinae, 170
Planigale, 12
Platacanthomyinae, 145–146
Platacanthomys, 145
Platalina, 50
Platanista, 89
Platanistidae, 89
Platyrrhinus, 52
Plecotus, 62
Plerotes, 38
Podogymnura, 23
Podomys, 156
Podoxymys, 156
Poecilogale, 82
Poelagus, 173
Pogonomelomys, 139
Pogonomys, 139
Poiana, 86
Pongo, 75
Pontoporia, 89
Potamochoerus, 92
Potamogale, 21
Potamogalinae, 21
Potoroidae, 15
Potorous, 15
Potos, 84
Potosinae, 84
Praomys, 139
Presbytis, 74
Primates, 66–75
Priodontes, 20
Prionailurus, 77
Prionodon, 86
Prionomys, 121
Proboscidea, 90
Procapra, 96–97
Procavia, 91
Procaviidae, 91
Procolobus, 74
Procyon, 84
Procyonidae, 84
Procyoninae, 84
Proechimys, 168–169
Proedromys, 119
Profelis, 77
Prolagus, 172

Prometheomys, 119
Promops, 66
Pronolagus, 173
Propithecus, 68
Prosciurillus, 104
Proteles, 80
Protelinae, 80
Protoxerus, 104
Psammomys, 124
Pseudalopex, 76
Pseudantechinus, 12
Pseudocheiridae, 17–18
Pseudocheirus, 17–18
Pseudochirops, 18
Pseudohydromys, 139
Pseudois, 99
Pseudomys, 139–140
Pseudorca, 88
Pseudoryzomys, 156
Ptenochirus, 38
Pteralopex, 38
Pteromyinae, 108–110
Pteromys, 110
Pteromyscus, 110
Pteronotus, 48
Pteronura, 80
Pteropodidae, 35–41
Pteropodinae, 35–40
Pteropus, 39–40
Ptilocercinae, 34
Ptilocercus, 34
Pudu, 95
Puertoricomys, 169
Puma, 78
Punomys, 156
Pygathrix, 74
Pygeretmus, 113–114
Pygoderma, 52

Quemisia, 171

Rangifer, 95
Raphicerus, 97
Rattus, 140–141
Ratufa, 104
Redunca, 100
Reduncinae, 100
Reithrodon, 157
Reithrodontomys, 157
Rhabdomys, 142
Rhagomys, 157
Rheithrosciurus, 104

Rheomys, 157
Rhinoceros, 91
Rhinocerotidae, 91
Rhinolophidae, 44–48
Rhinolophinae, 44–45
Rhinolophus, 44–45
Rhinonicteris, 48
Rhinophylla, 51
Rhinopoma, 41
Rhinopomatidae, 41
Rhinosciurus, 104
Rhipidomys, 157–158
Rhizomyinae, 146
Rhizomys, 146
Rhizoplagiodontia, 170
Rhogeessa, 62
Rhombomys, 124
Rhynchogale, 79
Rhyncholestes, 10
Rhynchomeles, 14
Rhynchomys, 142
Rhynchonycteris, 42
Rodentia, 100–171
Romerolagus, 173
Rousettus, 40
Rubrisciurus, 104
Rupricapra, 99
Ruwenzorisorex, 28

Saccolaimus, 42
Saccopteryx, 42–43
Saccostomus, 121
Saguinus, 69–70
Saiga, 97
Saimiri, 71
Salanoia, 78
Salpingotus, 114
Sarcophilus, 12
Scalopus, 33
Scandentia, 34
Scapanulus, 33
Scapanus, 33
Scapteromys, 158
Scaptochirus, 33
Scaptonyx, 33
Sciuridae, 100–110
Sciurillus, 104
Sciurinae, 100–108
Sciurotamias, 104
Sciurus, 104–105
Scleronycteris, 51
Scolomys, 158

Scotinomys, 158
Scotoecus, 62
Scotomanes, 62
Scotonycteris, 40
Scotophilus, 62
Scutisorex, 28
Sekeetamys, 124
Selevinia, 162
Semnopithecus, 74
Setifer, 21
Setonix, 17
Sicista, 114–115
Sicistinae, 114–115
Sigmoceros, 95
Sigmodon, 158
Sigmodontinae, 146–160
Sigmodontomys, 158
Sirenia, 90
Sminthopsis, 12–13
Solenodon, 20
Solenodontidae, 20
Solisorex, 28
Solomys, 142
Sorex, 30–32
Soricidae, 23–32
Soricinae, 29
Soriculus, 32
Sotalia, 88
Sousa, 88
Spaeronycteris, 52
Spalacinae, 160
Spalacopus, 166
Spalax, 160
Spelaeomys, 142
Speothos, 76
Spermophilopsis, 105
Spermophilus, 105–106
Sphaerias, 40
Sphiggurus, 163–164
Spilocuscus, 15
Spilogale, 81
Srilankamys, 142
Steatomys, 122
Stenella, 88
Steno, 88
Stenocephalemys, 142
Stenoderma, 52
Stenodermatinae, 51–53
Stenomys, 142
Stochomys, 142
Strigocuscus, 15
Sturnira, 53

Styloctenium, 40
Stylodipus, 114
Suidae, 92
Suinae, 92
Suncus, 28
Sundamys, 142
Sundasciurus, 106–107
Surdisorex, 28
Suricata, 79
Sus, 92
Syconycteris, 41
Sylvicapra, 99
Sylvilagus, 173
Sylvisorex, 28–29
Synaptomys, 119
Syncerus, 97
Syntheosciurus, 107

Tachyglossidae, 7
Tachyglossus, 7
Tachyoryctes, 146
Tadarida, 66
Taeromys, 142–143
Talpa, 33
Talpidae, 32–34
Talpinae, 32–34
Tamandua, 20
Tamias, 107
Tamiasciurus, 108
Tamiops, 108
Taphozous, 43
Tapiridae, 91
Tapirus, 91
Tarsiidae, 68–69
Tarsipedidae, 18
Tarsipes, 18
Tarsius, 68–69
Tarsomys, 143
Tasmacetus, 90
Tateomys, 143
Tatera, 124–125
Taterillus, 125
Taurotragus, 97
Taxidea, 82
Taxidiinae, 82
Tayassu, 92
Tayassuidae, 92
Tenrec, 21
Tenrecidae, 20–21
Tenrecinae, 21
Tetracerus, 97
Thallomys, 143

Thalpomys, 159
Thamnomys, 143
Theropithecus, 73
Thomasomys, 159
Thomomys, 111
Thoopterus, 40
Thrichomys, 169
Thryonomyidae, 163
Thryonomys, 163
Thylacinidae, 11
Thylacinus, 11
Thylamys, 10
Thylogale, 17
Thyroptera, 54
Thyropteridae, 54
Tokudaia, 143
Tolypeutes, 20
Tomopeas, 63
Tomopeatinae, 63
Tonatia, 49
Trachops, 49
Trachypithecus, 74
Tragelaphus, 97–98
Tragulidae, 93
Tragulus, 93
Tremarctos, 84
Triaenops, 48
Trichechidae, 90
Trichechus, 90
Trichosurus, 15
Trichys, 163
Trogopterus, 110
Tryphomys, 143
Tscherskia, 120

Tubulidentata, 91
Tupaia, 34
Tupaiidae, 34
Tupaiinae, 34
Tursiops, 88
Tylomys, 159–160
Tylonycteris, 62
Tympanoctomys, 167
Typhlomys, 146

Uncia, 78
Uranomys, 143
Urocyon, 76
Uroderma, 53
Urogale, 34
Uromys, 143
Uropsilinae, 34
Uropsilus, 34
Urotrichus, 34
Ursidae, 84–85
Ursinae, 84–85
Ursus, 85

Vampyressa, 53
Vampyrodes, 53
Vampyrum, 49
Vandeleuria, 143
Varecia, 67
Vernaya, 143
Vespertilio, 63
Vespertilionidae, 54–63
Vespertilioninae, 55–63
Vicugna, 93
Viverra, 86

Viverricula, 86
Viverridae, 85–86
Viverrinae, 86
Volemys, 119
Vombatidae, 14
Vombatus, 14
Vormela, 82
Vulpes, 76

Wallabia, 17
Wiedomys, 160
Wilfredomys, 160
Wyulda, 15

Xenarthra, 18–20
Xenomys, 160
Xenuromys, 144
Xeromys, 144
Xerus, 108

Zaedyus, 20
Zaglossus, 7
Zalophus, 83
Zapodinae, 115
Zapus, 115
Zelotomys, 144
Zenkerella, 161
Zenkerellinae, 160–161
Ziphiidae, 89–90
Ziphius, 90
Zygodontomys, 160
Zygogeomys, 111
Zyzomys, 144

INDEX TO COMMON NAMES

Aardvark, 91
Aardwolf, 80
Acouchis, 165
Addax, 99
Agoutis, 165
Alpaca, 93
Ammodile, 122
Angwantibos, 68
Anoas, 97
Anteaters, 20. *See also* Echid-
  nas; Pangolins
Antechinuses, 11
Antelopes, 95–98, 99–100
Aoudad, 98
Apes, 75
Aplodontia. *See* Mountain
  Beaver
Argali, 98
Armadillos, 19–20
Asses, 90–91
Aurochs, 97
Aye-Aye, 68

Babirusa, 92
Baboons, 73
Badgers, 80, 81, 82
Bandicoots, 13–14
Banteng, 97
Barasingha, 94
Barbastelles, 55
Bats, 35–66
    Abo, 55
    Allen's Striped, 55
    Bamboo, 62

Bats (*continued*)
  Banana, 51
  Barbastelles, 55
  Beatrix's, 55
  Bechstein's, 58
  Bent-winged, 63
  Bibundi, 55
  Big-eared, 62
    Allen's, 57
    Little, 48–49
    New Guinea, 60
  Big-eyed, 52
  Banford's, 56
  Blossom, 41
  Blunt-eared, 63
  Bonneted, 64
  Brandt's, 58
  Broad-eared, 66
  Broad-nosed, 52, 60
  Brown
    Argentine, 56
    Big, 56
    Big-eared, 57
    Brazilian, 56
    Little, 58
  Bulldog, 48
  Butterfly, 55
  Daubenton's, 58
  Dawn, 41
  Disk-footed, 56
  Disk-winged, 54
  Dog-faced, 64–65
  Dog-like, 42
  Epauletted, 37

Bats (*continued*)
  Evening, 57, 60
  False Vampire, 43
  Fig-eating, 51, 52
  Fish-eating, 59
  Flat-headed, 57, 65
  Flower, 50
  Flower-faced, 45
  Flying Foxes, 39–40
  Forest, 55, 56
  Free-tailed, 64–66
    African, 65
    Greater, 65
    Lesser, 64
    New World, 66
  Fringe-lipped, 49
  Fruit
    Black-bellied, 41
    Black-capped, 35
    Blanford's, 40
    Bulmer's, 35
    d'Anchieta's, 38
    Dobson's, 37
    Dwarf Epauletted, 37
    Dyak, 36
    Eidolon, 36
    Epauletted, 36
    Golden-capped, 35
    Hammer-headed, 37
    Harpy, 37
    Horsfield's, 35
    Island, 35
    Little, 51
    Little Collared, 37

Bats, Fruit (*continued*)
Long-tailed, 41
Long-tongued, 41
Lucas's Short-nosed,
38
Luzon, 38
Madagascan, 36
Manado, 35
Mindanao Pygmy, 35
Musky, 38
Naked-backed, 36
Old World, 35–41
Orange, 41
Palawan, 35
Panay Golden-
capped, 35
Pygmy, 35, 37
Ratanaworabhan's, 37
Red, 52
Rousette, 40
Salim Ali's, 37
Sao Tomé Collared, 37
Short-nosed, 35
Short-palated, 35
Small-toothed, 37
Spotted-winged, 35
Straw-colored, 36
Stripe-faced, 40
Sulawesi, 35
Sunda, 35
Swift, 40
Tailless, 37
Talaud, 35
Tube-nosed, 37–38
West African, 40
White-collared, 37
Woodford's, 41
Fruit-eating, 50, 51–52
Funnel-eared, 54
Geoffroy's, 58
Ghost, 42
Ghost-faced, 48
Golden, 49
Golden-tipped, 55
Great Stripe-faced, 53
Groove-toothed, 54
Hairless, 64
Hairy, 58, 59
Hairy-faced, 57
Hairy-nosed, 49
Hairy-tailed, 57
Hairy-winged, 63

Bats (*continued*)
Harlequin, 62
Heart-nosed, 43
Hoary, 57
Hodgson's, 58
Hog-nosed, 41
Horn-skinned, 56
Horseshoe, 44–45
Horsfield's, 58
House, 56, 62
Ikonnikov's, 58
Indiana, 59
Insectivorous Tube-
nosed, 63
Ipanema, 52
Kashmir Cave, 58
Large-footed, 57, 58
Leaf-chinned, 48
Leaf-nosed, 48
American, 48–54
Orange, 48
Tailless, 46
Trident, 46
Vietnam, 47
Little Brown, 57–59
Little Goblin, 65
Long-eared, 57, 60
Long-fingered, 58, 63
Long-legged, 48
Long-nosed, 51
Long-snouted, 50
Long-tailed, 50
Long-tongued, 50
Chestnut, 49
Dark, 51
Ega, 51
Mexican, 50
Underwood's, 50
MacConnell's, 52
Mastiff, 64, 65, 66
Monkey-faced, 38
Morris's, 59
Mouse-eared, 58, 59
Mouse-tailed, 41
Musky Fruit, 38
Mustached, 48
Myotis, 57–59
Naked-backed, 48
Natterer's, 59
Nectar, 49–50
New Zealand Short-
tailed, 64

Bats (*continued*)
Noctule, 59–60
Northern, 56
Nose-leaf, 48
Old World Fruit, 35–41
Painted, 55
Pale-faced, 49
Pallid, 55
Particolored, 63
Peter's Wrinkle-lipped,
65
Pied, 55
Pipistrelles, 60–61
Pond, 58
Pouched, 42
Proboscis, 42
Railer, 65
Red, 57
Rickett's Big-footed, 59
Ridley's, 59
Rohu's, 60
Round-eared, 49
Roundleaf, 46–47
Sac-winged, 41, 42–43
Schlieffen's, 60
Seminole, 57
Serotines, 55–56
False, 56–57
Shaggy, 41
Sheath-tailed, 41–43
Short-eared, 42
Short-tailed, 51, 64
Silver-haired, 57
Silvered, 55
Sind, 56
Single Leaf, 51
Slit-faced, 43
Smoky, 54
Sombre, 56
Spear-nosed, 49
Spectral, 49
Spotted, 56
Sucker-footed, 54
Sword-nosed, 48
Tacarcuna, 57
Tailless, 50
Tent-making, 53
Thick-eared, 56
Thick-thumbed, 56
Thumbless, 54
Tickell's, 57
Tomb, 43

Bats (continued)
Tree, 51
Trident, 46, 48
Trident-nosed, 46
Trumpet-eared, 54, 55
Tube-nosed, 38, 63
Vampire, 53–54
False, 43
Van Gelder's, 55
Veldkamp's, 37
Vesper, 54–63
Visored, 52
Wattled, 55
Welwitch's, 59
Whiskered, 59
White, 52
White-shouldered, 51
Woermann's, 41
Woolly, 48, 54–55
Wrinkle-faced, 52
Yellow, 57, 62
Yellow-eared, 53
Yellow-lipped, 56
Yellow-shouldered, 53
Yellow-winged, 43
Bears, 84–85
Beavers, 110
Mountain, 100
Beira, 96
Beluga, 88
Bettongs, 15
Bharals, 99
Bilbies, 13
Bilby, 13
Binturong, 85
Bison, 97
Blackbuck, 96
Blesmols, 162
Blue Buck, 99
Boars, 92
Bobcat, 77
Bongo, 97
Bontebok, 95
Brockets, 94–95
Buffaloes, 97
Bushbuck, 97

Cacomistle, 84
Camels, 93
Capuchins, 71
Capybara, 165
Caracal, 77

Caribou, 95
Carnivores, 75–86
Cats, 76–78
Cattle, 97
Cavies, 164
Chamois, 99
Cheetah, 76
Chevrotains, 93
Chimpanzees, 75
Chinchillas, 164
Chipmunks, 107
Chiru, 96
Chital, 93
Civets, 85–86
Coatis, 84
Coendus. See Porcupines,
New World
Coruro, 166
Cottontails, 173
Cougar. See Puma
Coyote, 75
Culpeo, 76
Cuscuses, 14–15
Cusimanses, 78

Dasyures, 11
Dasyurids, 11–13
Deer, 93–95
Degus, 166
Desmans, 32
Dhole, 76
Dibatag, 95
Dibblers, 12
Dik-diks, 96
Dogs, 75–76
Prairie, 101
Dolphins
Marine, 87–88
River, 89
Dorcopsises, 16
Dormice, 161–162
Chinese Pygmy, 146
Malabar Spiny, 145
Drill, 73
Dromedary, 93
Dugong, 90
Duikers, 99
Dunnarts, 12–13

Echidnas, 7
Echymiperas, 13
Edentates, 18–20

Elands, 97
Elephant Shrews, 173–174
Elephants, 90
Elk, 94
Ermine, 82

Falanouc, 85
False Serotines, 56–57
Fennec, 76
Ferret, 82. See also Polecat,
European
Fisher, 81
Flying Foxes, 39–40
Flying Lemurs, 35
Fossa, 85
Foxes, 75, 76
Franciscana, 89

Galagos, 68
Gaur, 97
Gazelles, 96–97
Gemsbok, 99
Genets, 86
Gerbils, 122–124, 122–125
Cape Short-eared, 122
Fat-tailed, 124
Great, 124
Hairy-footed, 122
Large Naked-soled,
124–125
Mongolian. See Jirds,
Mongolian
Pouched, 122
Przewalski's, 122
Small Naked-soled, 125
Somali Pygmy, 124
Gerenuk, 96
Gibbons, 75
Giraffe, 93
Gliders, 17, 18
Goats, 98
Golden Moles, 21–22
Gophers, Pocket, 110–111
Gorals, 98
Gorilla, 75
Grisons, 81
Groundhog. See Woodchuck
Grysboks, 97
Guanaco, 93
Guemals, 94
Guenons, 72–73
Guereza, 73

Guiara, 168
Guinea Pigs, 164
Gundis, 161
Gymnures, 23

Hamsters, 120
Hare, Spring, 161
Hare-wallabies, 16
Hares, 172
Hartebeest, 95
Hedgehogs, 22–23
Hippopotamuses, 92–93
Hirola, 95
Hocicudos, 154
Hogs, 92
Horses, 90–91
Humans, 75
Hutias, 170
Hyenas, 79–80
Hyraxes, 91

Ibexes, 98
Impala, 95
Indri, 67
Insectivores, 20–34

Jackals, 75
Jackrabbits, 172
Jaguar, 78
Jaguarundi, 77
Jerboas, 113–114
Jirds, 124

Kaluta, Little Red, 11
Kangaroos, 15–16
Kiang, 91
Kinkajou, 84
Klipspringer, 96
Koalas, 14
Kobs, 100
Kodkod, 77
Kouprey, 97
Kowari, 11
Kudus, 97, 98
Kulan, 91
Kultarr, 12

Langurs, 74
Lechwes, 100
Lemmings
    Amur, 117
    Arctic, 116
    Bog, 119

Lemmings (continued)
    Brown, 117
    Collared, 116
    Mole, 116–117
    Norway, 117
    Steppe, 117
    Wood, 119
    Wrangel, 116
Lemurs, 66–68
    Flying, 35
Leopards, 78
Linsangs, 86
Lion, 78
    Mountain. See Puma
Lion Tamarins, 69
Llama, 93
Lorises, 68
Lynxes, 77

Macaques, 73
Macropods, 15–17
Manatees, 90
Mandrill, 73
Mangabeys, 72, 73
Maras, 164
Margay, 77
Markhor, 98
Marmosets, 69
Marmots, 103
Marsupial Mice, 11
Marsupial Moles, 14
Marsupial Shrews, 12
Marsupials, Carnivorous,
    11–13
Martens, 81
Meerkat, 79
Mice
    African, 136–137
    African Soft-furred, 139
    African Wood, 131
    Algerian, 136
    Alice Springs, 140
    Altiplano, 149
    Andean, 147
    Angel Island, 155
    Ash-gray, 139
    Australian, 139–140
    Australian Hopping,
        137–138
    Australian Native, 131
    Aztec, 154
    Baoule's, 135

Mice (continued)
    Beady-eyed, 159
    Big-eared, 148
    Big-footed, 145
    Birch, 114–115
    Black-eared, 155
    Black-tailed, 155
    Blue-gray, 140
    Blunt-toothed, 171
    Bolam's, 139
    Bolo, 148
    Brazilian Arboreal, 157
    Bristly, 151
    Broad-headed, 144
    Broad-toothed, 140
    Brown, 158
    Brush, 128, 139, 154
    Cactus, 155
    California, 155
    Callewaert's, 136
    Canary, 128
    Candango, 150
    Cane, 160
    Canyon, 155
    Cerrado, 159
    Chaco, 147
    Chestnut, 140
    Chihuahuan, 155
    Chinchilla, 149
    Climbing
        African, 121
        American, 157–158
        Chilean, 150
        Long-tailed, 143
        Red, 143
        Velvet, 121
    Columbian, 155
    Cook's, 136
    Cotton, 155
    Country, 140
    Crump's, 129
    Dalton's, 137
    Deer, 154–156
        Crested-tailed, 149
        Giant, 150
        Michoacan, 154
    Delany's Swamp, 145
    Delicate, 136
    Deroo's, 137
    Desert, 140, 155
    Earth-colored, 136
    False Canyon, 155

Mice (*continued*)
Fat, 122
Fawn-colored, 136
Field, 126, 135. *See also*
  Voles, Meadow
Flat-haired, 136
Florida, 156
Forest
  Colombian, 149
  Congo, 121
  Groove-toothed, 121
  Philippine, 126–127
Galapagos, 152
Garlepp's, 149
Gerbil, 121, 149
Gleaning, 156
Golden, 152
Gould's, 140
Gounda, 136
Grass, 146–147
  Four-striped, 142
  Striped, 131–132
Grasshopper, 153
Harvest
  American, 157
  Eurasian, 135
Hastings River, 140
Hausa, 136
Hooper's, 155
Hopping, 137–138
House, 136
Inland Sandy, 140
Island, 145
Jumping, 115, 133
Juscelin's, 150
Kangaroo, 112
Kasai, 136
Key, 170–171
Kimberly, 140
Lava, 133
Leaf-eared, 149, 156
Long-clawed, 148, 152
Long-clawed Mole,
  149
Long-tailed, 140
Macedonian, 136
Mahomet, 136
Marsh, 155
Marsupial, 14
Matthey's, 136
Maya, 155
Mayor's, 136

Mice (*continued*)
Meadow. *See* Voles,
  Meadow
Mesquite, 155
Molelike, 150
Montane, 146
Mound-building, 136
Multimammate, 133–134
Native, 131, 139
Nayarit, 155
Neave's, 136
New Holland, 140
Nikolaus's, 121
Nimble-footed, 155
Old World, 135–136
Oldfield, 155, 159
Orange, 136
Osgood's, 155
Oubangui, 136
Pebble-mound, 139, 140
Perote, 154
Peter's, 136
Phillips's, 136
Pilliga, 140
Pinyon, 156
Plains, 139
Plate-toothed, 171
Plateau, 155
Pocket, 112–113
Pouched, 121
Puna, 156
Pygmy, 136
  American, 148
  Desert, 136
  Gray-bellied, 136
  Setzer's, 136
  Thomas's, 136
Ranee, 130
Red-nosed, 160
Rock, 137, 145, 155
Rock-loving, 136
Roraima, 156
Rudd's, 143
Rupp's, 137
Ryukyu, 136
San Esteban Island,
  156
San Lorenzo, 155
Santa Cruz, 155
Servant, 136
Shark Bay, 140
Shortridge's, 136

Mice (*continued*)
Shrew
  Brazilian, 148
  Lesser, 135
  Mottled-tailed, 137
  Mt. Isarog, 127
  New Guinean, 139
  One-toothed, 134
Shrewmouse
  Gairdner's, 136
  Silky, 139
Sitka, 156
Slevin's, 156
Smoky, 130, 140
Spiny, 158
  African, 125
  Ceylon, 136
  Striped, 130
  Striped-back, 135
  Sumatran Shrewlike, 136
  Temminck's, 136
  Texas, 154
  Toad, 135
  Tree, 128
    Dollman's, 121
    Pencil-tailed, 128
    Prehensile-tailed, 139
  Tres Marias Island, 155
  Twisted-toothed, 171
  Verreaux's, 137
  Vesper, 148
  Volcano, 136, 151
  Water, 149, 157
  Western, 140
  White-ankled, 155
  White-footed, 136, 155
  White-tailed, 144
  Wilfred's, 160
  Winkelmann's, 156
  Yemeni, 137
Mink, 82
Mole Rats, 160, 162
Moles, 32–34
  East Asian, 33
  Eastern, 33
  Gansu, 33
  Golden, 21–22
  Long-tailed, 33
  Marsupial, 14
  Old World, 33
  Oriental, 32–33
  Ryukyu, 33

Moles (*continued*)
  Short-faced, 33
  Shrew, 33, 34
  Star-nosed, 32
  Western American, 33
  White-tailed, 33
Mongooses, 78–79
Monito Del Monte, 10
Monjon, 17
Monkeys
  Allen's Swamp, 72
  Capuchin, 71
  Colobus, 73–74
  Goeldi's, 69
  Guenons, 72–73
  Guerezas, 73
  Howler, 70
  Langurs, 74
  Leaf, 74
  Long-nosed, 74
  Mangabeys, 72, 73
  Muriquis, 71
  New World, 70–72
  Night, 70
  Old World, 72–74
  Patas, 73
  Proboscis, 74
  Rhesus, 73
  Sakis, 72
  Snub-nosed, 74
  Spider, 70–71
  Squirrel, 71–72
  Talapoins, 73
  Titis, 71
  Uakaris, 72
  Vervet, 73
Monotremes, 7
Moonrat, 23
Moose, 94
Mouflon, 98
Mountain Beaver, 100
Mountain Goat, 98
Mouse-deer, 93
Mulgara, 11
Muntjacs, 94
Muriqui, 71
Muskox, 98
Muskrats, 119
Myotis, 57–59

Narwhal, 88
Nesophontes, 20

Nilgai, 97
Ningauis, 12
Numbat, 11
Nutria, 171
Nyalas, 97

Ocelot, 77
Octodonts, 166–167
Okapi, 93
Olingos, 84
Onager, 91
Opossums
  American, 7–10
  Big-eared, 7
  Black-shouldered, 7
  Bushy-tailed, 7
  Fat-tailed, 10
  Four-eyed, 9, 10
  Gracile Mouse, 8
  Large American, 7–8
  Lutrine, 8
  Mouse, 8–9
  Patagonian, 8
  Short-tailed, 9–10
  Shrew, 10
  Slender Mouse, 8–9
  Southern, 8
  Virginia, 8
  Water, 7
  White-eared, 7
  Woolly, 7
  Woolly Mouse, 9
Orangutans, 75
Oribi, 96
Oryxes, 99
Otter Shrews, 21
Otters, 80
Oxen, 97

Pacarana, 164
Pacas, 165
Pademelons, 17
Pandas, 84
Pangolins, 100
Peccaries, 92
Pectinator, Speke's, 161
Phascogales, 12
Pichi, 20
Pigs, 92
Pikas, 171–172
Pipistrelles, 60–61
Planigales, 12

Platypus, 7
Pocket Gophers, 110–111
Polecats, 81, 82
Porcupines
  New World, 163–164
  Old World, 162–163
Porpoises, 88–89
Possums
  Brushtail, 15
  Feathertail, 18
  Gliding, 18
  Honey, 18
  Leadbeater's, 18
  Pygmy, 17
  Ringtail, 17–18
  Scaly-tailed, 15
  Striped, 18
Potoroos, 15
Potto, 68
Prairie Dogs, 101
Primates, 66–75
Pronghorn, 95
Pseudantechinuses, 12
Pudus, 95
Puku, 100
Puma, 77
Punare, 169

Quagga, 91
Quokka, 17
Quolls, 11

Rabbits, 172, 173
Raccoons, 84
Rat Kangaroos, 15
Rats
  Acacia, 143
  African Grass, 127
  African Groove-toothed,
    136
  Andaman, 141
  Andean, 150
  Annandale's, 140
  Arboreal, 156
  Armored, 168
  Australian Stick-
    nest, 132
  Bagobo, 127
  Bamboo, 146, 167
  Bandicoot, 127, 137
  Bartel's, 142

Rats *(continued)*

Black, 141. *See also* Rats, House

Blanford's, 129

Bonthain, 140

Bristle-spined, 167

Brown, 141

Brush-furred, 132

Brush-tailed, 168

Buhler's, 129

Bulldog, 141

Bunny, 157

Bush, 130, 140

Cane, 163

Cape York, 141

Cave, 142, 169

Celebes, 142

Ceram, 140, 142

Chinchilla, 167

Climbing, 154, 159–160

Cloud, 129, 138–139

Corozal, 169

Cotton, 158

Crab-eating, 150

Crested, 125

Crimson-nosed, 148

Cutch, 129

Dassie, 163

Defua, 129

Dusky, 140

Eastern, 141

Edible, 169

Elvira, 129

Emperor, 143

Enggano, 140

Ethiopian Narrow-headed, 142

Field, 141

Fish-eating, 148, 152

Flores Long-nosed, 138

Forest

Atlantic, 149

Luzon, 127

Philippine, 140

Sulawesi, 143

Tawi-tawi, 141

Gambian, 121

Giant, 121

Bismarck, 143

Long-tailed, 132

Malagasy, 145

Mountain, 142

Rats, Giant *(continued)*

New Guinean, 131

Rock-dwelling, 144

South American, 150

Sulawesi, 138

Trefoil-toothed, 132

Giant Sunda, 142

Giluwe, 140

Glacier, 142

Guadalcanal, 143

Hainald's, 140

Hairy-tailed, 127

Harrington's, 129

Heath, 140

Hill, 127–128

Hoffman's, 140

Hole, 140

Hoogerwerf's, 140

House, 141

Indian, 129

Isarog Striped Shrew-rat, 128

Isthmus, 150

Ivory Coast, 129

Japen, 141

Kangaroo, 111–112

Kerala, 141

King, 143

Komodo, 131

Koopman's, 141

Korinch's, 141

Limestone, 137

Long-footed, 143

Long-haired, 141

Loring's, 143

Lovely-haired, 142

Luzon Broad-toothed, 125

Luzon Short-nosed, 143

MacLear's, 141

Magdalena, 160

Margareta's, 133

Marmoset, 130

Marsh, 129, 150

Mentawai, 141

Miliard's, 129

Mindoro, 126

Molaccan Prehensile-tailed, 141

Mole, 146, 160, 162

Mosaic-tailed, 134–135

Moss-forest, 142

Rats *(continued)*

Mountain, 132, 137, 138

Mt. Oku, 131

Naked-tailed, 142, 143

New Guinean, 133, 141

New Ireland, 141

Nillu, 141

Nonsense, 140

Norway. *See* Rats, Brown

Ohiya, 142

Olalla, 167

Old World, 140–141

Opossum, 141

Osgood's, 141

Pack. *See* Woodrats

Palm, 141

Peleng, 141

Philippine, 127

Philippine Striped, 128

Polynesian, 140

Pouched, 121

Rabbit, 129

Rice, 153

Arboreal, 152

Brazilian False, 156

Dark, 150

Dusky, 150

Pygmy, 152–153

Small, 150

West Indian Giant, 150

Rice-field, 140, 141

Rock, 166

African, 125–126

Australian, 144

Chilean, 166

Sage's, 166

Rufous-nosed, 138

Ruschi's, 146

Ryukyu, 129

Salokko, 142

Sand, 124

Shaggy, 129

Short-tailed, 144

Shortridge's, 143

Shrew, 142

Greater Sulawesian, 143

Isarog Striped, 128

Luzon Blazed, 128

Philippine, 129

Sulawesian, 134

Sikkim, 141

Rats (continued)
  Simalur, 141
  Slender, 142
  Small-toothed, 133
  Smoke-bellied, 137
  Soft-furred, 130, 135, 141
  Spiny, 141
    American, 167–169
    Arboreal Soft-furred,
      167
    Armored, 168
    Bishop's Fossorial, 168
    Broad-headed, 168
    Lund's, 168
    Oriental, 134
    Owl's, 168
    Ryukyu, 143
    Sulawesi, 130
    Terrestrial,168–169
  Squirrel-toothed, 126
  Stein's, 141
  Sula, 140
  Sulawesi, 142–143
  Sulawesi Bear, 138
  Sulawesi Montane, 142
  Summit, 140
  Sunburned, 140
  Swamp, 158
    African, 133
    Andean, 151
    Australian, 141
    Groove-toothed, 138
  Tanezumi, 141
  Target, 142
  Thicket, 130, 143
  Timor, 141
  Tondano, 143
  Tree, 135
    Black-tailed, 143
    Dark-tailed, 137
    Fea's, 128
    Flores Island Giant,
      138
    Gray, 132
    Luzon, 128
    Red Crested, 167
    Rufous, 167
    Sody's, 131
    Spiny, 167–168, 168
    Sunda, 139
  Tufted-tailed, 144–145
  Turkestan, 141

Rats (continued)
  Van Deusen's, 142
  Vesper, 152, 154
  Viscacha, 166, 167
  Vlei, 145
  Water, 130–131
    African, 128
    Coarse-haired, 138
    Earless, 129
    False, 144
    Montane, 138
    Neotropical, 151
    New Guinean, 132
    Rice, 158
  Whistling, 145
  White-bellied, 137
  White-tailed, 143, 144
  White-toothed, 127
  Wood. See Woodrats
  Woolly, 133
  Yellow-tailed, 141
Reedbucks, 100
Reindeer. See Caribou
Rhebok, Common, 100
Rhinoceroses, 91
Ringtails, 17–18, 84
Rockhares, 173
Rodents, 100–171
Rorquals, 86–87

Sable, 81
Saiga, 97
Sakis, 72
Sambar, 94
Sand Rats, 124
Sassabies, 95
Scaly-tail, Cameroon, 161
Sea Cow, Steller's, 90
Sea Lions, 82–83
Seals
    Eared. See Sea Lions
    Earless, 83
    Fur, 82
Serotines, 55–56
    False, 56–57
Serows, 98
Serval, 77
Sheep, 98
Shrew-rat, Isarog Striped, 128
Shrewmouse, Gairdner's, 136
Shrews, 23–32
  Aberdare, 28

Shrews (continued)
  African, 28
  African Foggy, 23
  Alexandrian, 23
  Alpine, 30
  Amami, 26
  American Short-tailed,
    29
  Andaman, 23
  Andaman Spiny, 24
  Anderson's, 28
  Ansell's, 23
  Apennine, 31
  Arabian, 23
  Arctic, 30
  Arizona, 30
  Armenian, 23
  Armored, 28
  Arrogant, 28
  Asian Short-tailed, 29
  Asiatic, 32
  Azumi, 31
  Bailey's, 23
  Baird's, 30
  Bale, 23
  Barren Ground, 32
  Bates's, 23
  Beccari's, 23
  Bicolored, 25
  Black, 28
  Black-footed, 26
  Bottego's, 23
  Brown-toothed, 32
  Buettikofer's, 23
  Butiaba Naked-tailed, 25
  Canary, 23
  Carmen Mountain, 31
  Caucasian, 31
  Celebes, 25
  Chestnut-bellied, 32
  Chinese, 31
  Cinderella, 24
  Cinereus, 30
  Climbing, 28
  Congo, 24
  Crescent, 28
  Crosse's, 24
  Crowned, 30
  Dark, 25
  Day's, 28
  De Winton's, 32
  Dent's, 24

Shrews (continued)

Desert, 30
Desperate, 24
Dhofarian, 24
Dramatic, 25
Dsinezumi, 24
Dusky, 26
Dwarf, 28, 31
Eisentraut's, 24
Elephant, 173–174
Elgon, 24
Elongated, 24
Eurasian, 30
Even-toothed, 31
Fischer's, 24
Flat-headed, 26
Flat-skulled, 31
Flores, 28
Flower's, 24
Fog, 31
Forest, 27
Fox's, 24
Fuscous, 26
Gansu, 30
Gaspé, 31
Glass's, 24
Goliath, 24
Grant's, 28
Grasse's, 24
Grauer's, 28
Greater, 28
Greenwood's, 24
Gueldenstaedt's, 24
Guramba, 26
Harenna, 24
Heather, 24
Highland, 23
Hildegarde's, 24
Himalayan, 32
Holarctic, 30–32
Horsfield's, 24
Hose's, 28
House, Asian, 28
Howell's, 28
Hun, 23
Indochinese, 23
Inyo, 32
Iranian, 27
Isabella, 28
Jackson's, 24
Jenkin's, 25
Johnston's, 28

Shrews (continued)

Jungle, 28
Kamchatka, 30
Kashmir, 31
Kenyan, 28
Kivu, 25
Kozlov's, 31
Lagranja, 31
Lamotte's, 25
Lamulate, 32
Large-headed, 24
Large-toothed, 25, 31
Latona, 25
Laxmann's, 30
Least, 30
Lemara, 25
Lesser, 26
Lofty, 30
Long-clawed, 27, 28, 32
Long-footed, 24
Long-tailed, 25, 30, 31
Lowe's, 32
Luzon, 24
MacArthur's, 25
MacMillan's, 25
Macow's, 25
Madagascan, 28
Malayan, 25
Manenguba, 25
Marsh, 30
Marsupial, 12
Mauritanian, 25
Max's, 25
Merriam's, 31
Mexican, 30
Mindanao, 23
Mindoro, 25
Miniscule, 31
Minute, 25
Mole, 29
Montane, 31
Moon, 26
Moorland, 25
Mountain, 29, 32
Mouse, 27
Mouse-tailed, 25
Mt. Lyell, 31
Mt. Malindang, 24
Musk, 24–26, 28–29
Narrow-headed, 26
Neglected, 25
Negros, 26

Shrews (continued)

Nicobar, 26
Nigerian, 26
Nimba, 26
Olivier's, 26
Ornate, 31
Osorio, 26
Pacific, 31
Palawan, 26
Pale Gray, 26
Pamir, 30
Pantellerian, 24
Paradox, 26
Paramushir, 31
Pasha, 26
Piebald, 27
Pitch, 26
Pitman's, 26
Poll's, 23
Portenko's, 31
Prairie, 31
Preble's, 31
Pribilof Island, 31
Pygmy, 26, 28, 31, 32
Radde's, 31
Rainey, 26
Remy's, 28
Rombo, 25
Roosevelt's, 26
Ruwenzori, 28
Sado, 31
Salenski's, 32
San Cristobal, 32
Sao Tomé, 27
Saussure's, 31
Savanna, 24
Savanna Path, 27
Savanna Swamp, 25
Schouteden's, 28
Sclater's, 31
Serezkaya, 26
Shinto, 31
Short-tailed, 29
Sicilian, 26
Slender, 31
Small-eared, 29–30
Small-footed, 26
Smith's, 32
Smoky, 30
Somali, 26
Southeast Asian, 24
Southeastern, 31

Shrews (continued)
Sri Lanka, 28
Sri Lanka Highland, 28
St. Lawrence Island, 31
Stony, 26
Stripe-backed, 30
Striped, 30
Sulawesi, 25
Sunda, 25
Tanzanian, 27
Tarfaya, 27
Telford's, 27
Temboan, 26
Tenebrous, 26
Thalia, 27
Therese's, 27
Thin, 27
Tibetan, 32
Tien Shan, 30
Tree, 34
Trowbridge's, 32
Tumultuous, 27
Tundra, 32
Ugandan, 27
Ultimate, 27
Usambara, 27
Ussuri, 31
Vagrant, 32
Verapaz, 32
Vermiculate, 27
Voi, 27
Volcano, 29
Water, 31
    Elegant, 30
    Glacier Bay, 30
    Old World, 30
    Oriental, 29
Whitaker's, 27
White-toothed, 23–27,
    26
    Black, 26
    Dwarf, 25
    Montane, 25
    Obscure, 26
    Smoky, 24
    Tiny, 25
    Ussuri, 25
Wimmer's, 27
Yankari, 27
Zacatecas, 30
Zaphir's, 27
Zarudny's, 27

Shrews (continued)
Zimmermann's, 27
Zimmer's, 27
Siamang, 75
Sifakas, 68
Sitatunga, 98
Skunks, 81
Sloths, 18–19
Solenodons, 20
Spring Hare, 161
Springbok, 95
Squirrels, 100–110
    Abert's, 104
    Allen's, 104
    Andean, 105
    Anderson's, 101
    Antelope, 100
    Black-banded, Borneo,
        101
    Black-eared, 103
    Black-striped, 101
    Bolivian, 105
    Brooke's, 106
    Bush, 103–104
    Busuanga, 106
    Caucasian, 104
    Collie's, 104
    Davao, 106
    Deppe's, 104
    Douglas's, 108
    Dwarf, 103, 104
    Ear-spot, 101
    Fiery, 104
    Finlayson's, 101
    Flying, 108–110
        Arrow-tailed, 108–109
        Black, 108
        Complex-toothed, 110
        Dwarf, 109
        Eurasian, 110
        Giant, 109
        Hairy-footed, 108
        Horsfield's, 109
        Namdapha, 108
        New World, 108
        North Chinese, 108
        Pygmy, 109
        Scaly-tailed, 160
        Smoky, 110
        Thomas's, 108
        Woolly, 108
    Fox, 105

Squirrels (continued)
    Fraternal, 106
    Giant, 104
    Gray, 104, 105
    Gray-bellied, 101
    Ground, 105–106
        African, 108
        Barbary, 101
        Bornean Mountain,
            101
        Indochinese, 103
        Long-clawed, 105
        Niobe, 103
        Striped, 102–103
        Tufted, 104
    Guayaquil, 105
    Guianan, 104
    Himalayan, Orange-
        bellied, 101
    Horse-tailed, 106
    Inornate, 101
    Irrawaddy, 101
    Japanese, 105
    Jentink's, 107
    Kinabalu, 101
    Kloss, 101
    Long-nosed, 101, 102
    Low's, 107
    Mearns's, 108
    Mentawai, 101
    Mentawai Three-
        striped, 103
    Mindanao, 107
    Montane, 107
    Mountain, 102, 103, 107
    Pallas's, 101
    Palm, 101, 102
    Peters's, 105
    Phayre's, 101
    Plantain, 101
    Prevost's, 101
    Pygmy, 101–102, 103, 104
    Red, 105, 108
    Red-bellied, 104
    Red-cheeked, 101
    Red-hipped, 101
    Red-tailed, 105
    Richmond's, 105
    Rock, 104, 105, 106
    Rope, 102
    Samar, 107
    Sanborn's, 105

Squirrels (*continued*)
 Scaly-tailed, 160
 Sculptor, 102
 Shrew-faced, 104
 Slender, 107
 Slender-tailed, 104
 Striped, Asiatic, 108
 Sun, 102
 Sunda, 106–107
 Tree, 101, 104–105, 107
 Variegated, 105
 Yellow-throated, 104
 Yucatan, 105
Steenbok, 97
Suni, 96

Tahrs, 98
Takin, 98
Talapoin, 73
Tamanduas, 20
Tamaraw, 97
Tamarins, 69–70
Tapeti, 173
Tapirs, 91
Tarsiers, 68–69
Tasmanian Devil, 12
Tasmanian Wolf. *See* Thylacine
Tayra, 81
Tenrecs, 20–21
Thylacine, 11
Tiger, 78
Titis, 71
Topi, 95
Tree Shrews, 34
Trioks, 18
Tuco-tucos, 165–166
Turs, 98

Uakaris, 72
Ungulates
 Even-toed, 92–100
 Odd-toed, 90–91
Urial, 98

Vaquita, 89
Vicugna, 93
Viscachas, 164
Voalavoanala, 145
Voles
 Afghan, 116
 Altai, 118

Voles (*continued*)
 Baluchistan, 118
 Bank, 116
 Beach, 117
 Blyth's, 118
 Brandt's, 117
 Bucharian, 116
 Cabrera's, 117
 California, 117
 Central Kashmir, 115
 Chaotung, 117
 Chinese Scrub, 118
 Clarke's, 119
 Common, 117
 Creeping, 118
 Duke of Bedford's, 119
 Evorsk, 118
 Felten's, 118
 Field, 117
 Flat-headed, 115
 Ganzu, 117
 Gerbe's, 118
 Gray-tailed, 118
 Guatemalan, 118
 Günther's, 118
 Heather, 119
 Insular, 117
 Japanese, 119
 Japanese Grass, 118
 Juniper, 118
 Kashmir, 117
 Kolan, 117
 Lacustrine, 118
 Large-eared, 115
 Lemming, 115
 Long-clawed Mole, 119
 Long-tailed, 118
 Mandarin, 117
 Marie's, 119
 Maximowicz's, 118
 Meadow, 117–119, 118
 Mexican, 118
 Middendorf's, 118
 Mongolian, 118
 Montane, 118
 Mountain, 115
 Muisk, 118
 Murree, 117
 Musser's, 119
 Narrow-headed, 118
 Nasarov's, 118

Voles (*continued*)
 Père David's, 117
 Persian, 118
 Pine, 117, 118
 Plateau, 117
 Prairie, 118
 Pratt's, 117
 Red-backed, 116
 Reed, 118
 Rock, 118
 Royal, 117
 Sagebrush, 117
 Sakhalin, 119
 Shansei, 117
 Shikotan, 116
 Siberian, 118
 Sikkim, 119
 Silver, 115
 Singing, 118
 Smith's, 119
 Snow, 116
 Social, 119
 South Asian, 117
 Southern, 119
 Southwest China, 117
 Szechuan, 119
 Taiga, 119
 Taiwan, 119
 Tarabundi, 118
 Tien Shan, 118
 Townsend's, 119
 Transcaspian, 119
 Tree, 115
 True's, 117
 Tundra, 118
 Water, 116, 119
 White-footed, 115
 Woodland, 118
 Yulungshan, 117
 Zempoaltepec, 119

Wallabies, 16–17
Wallaroos, 16
Walrus, 82
Warthogs, 92
Waterbuck, 100
Weasels, 81, 82
Whales
 Beaked, 89–90
 Bottlenose, 89
 Bowhead, 86
 Killer, 86, 87

Whales (*continued*)
    Right, 86, 87
    Rorquals, 86–87
    Sperm, 89
    Strap-toothed, 90
    White, 88

Wildebeests, 95
Wolverine, 81
Wolves, 75, 76
    Tasmanian. *See* Thy-
      lacine
Wombats, 14

Woodchuck, 103
Woodrats, 149, 151

Yak, 97

Zebras, 90, 91
Zokors, 144